THE
EPIGENETICS
REVOLUTION

THE
EPIGENETICS
REVOLUTION

NESSA CAREY

How Modern Biology
is Rewriting Our Understanding
of Genetics, Disease
and Inheritance

ICON

This edition published in the UK in 2012 by
Icon Books Ltd, Omnibus Business Centre,
39–41 North Road, London N7 9DP
email: info@iconbooks.com
www.iconbooks.com

Previously published in the UK in 2011 by Icon Books Ltd

Sold in the UK, Europe and Asia
by Faber & Faber Ltd, Bloomsbury House, 74–77 Great Russell Street,
London WC1B 3DA or their agents

Distributed in the UK, Europe and Asia
by TBS Ltd, TBS Distribution Centre, Colchester Road
Frating Green, Colchester CO7 7DW

Distributed in Australia and New Zealand
by Allen & Unwin Pty Ltd, PO Box 8500,
83 Alexander Street, Crows Nest, NSW 2065

Distributed in South Africa by Jonathan Ball,
Office B4, The District, 41 Sir Lowry Road,
Woodstock 7925

ISBN: 978-184831-347-7

Typeset in 12 on 16pt Times by Marie Doherty

Printed and bound in the UK by
Clays Ltd, St Ives plc

For Abi Reynolds, who reprogrammed my life
And in memory of Sean Carey, 1925 to 2011

Nessa Carey has a virology PhD from the University of
Edinburgh and is a former Senior Lecturer in Molecular Biology
at Imperial College London. She has worked in the biotech and
pharmaceutical industry for ten years. She lives in Bedfordshire
and this is her first book.

www.nessacarey.co.uk
@NessaCarey

Contents

Acknowledgements

Over the last few years I've had the privilege of working with some truly amazing scientists. There are too many to name here but special acknowledgements must go to Michelle Barton, Stephan Beck, Mark Bedford, Shelley Berger, Adrian Bird, Chris Boshoff, Sharon Dent, Didier Devys, Luciano Di Croce, Anne Ferguson-Smith, Jean-Pierre Issa, Peter Jones, Bob Kingston, Tony Kouzarides, Peter Laird, Jeannie Lee, Danesh Moazed, Steve McMahon, Wolf Reik, Ramin Shiekhattar, Irina Stancheva, Azim Surani, Laszlo Tora, Bryan Turner and Patrick Varga-Weisz.

Thanks go also to my former colleagues at CellCentric – Jonathan Best, Devanand Crease, Tim Fell, David Knowles, Neil Pegg, Thea Stanway and Will West.

As a first-time author I owe major gratitude to my agent, Andrew Lownie, for taking a risk on me and on this book.

Major thanks also to the lovely people at my publishers Icon, especially Simon Flynn, Najma Finlay, Andrew Furlow, Nick Halliday and Harry Scoble. Their unfailing patience with my complete ignorance of all aspects of publishing has been heroic.

I've had great support from family and friends and I hope they'll forgive me for not mentioning them all by name. But for sheer entertainment and distraction during some stressy patches I have to thank Eleanor Flowerday, Willem Flowerday, Alex Gibbs, Ella Gibbs, Jessica Shayle O'Toole, Lili Sutton and Luke Sutton.

And for always resisting the temptation to roll her eyes every time I said, 'I can't meet friends/do the dishes/go away for the weekend because I'm working on my book', I've got to thank my lovely partner Abi Reynolds. I promise I'll take that ballroom dancing lesson now.

Introduction

DNA.

Sometimes, when we read about biology, we could be forgiven for thinking that those three letters explain everything. Here, for example, are just a few of the statements made on 26 June 2000, when researchers announced that the human genome had been sequenced[1]:

> *Today we are learning the language in which God created life.*
> US President Bill Clinton

> *We now have the possibility of achieving all we ever hoped for from medicine.*
> UK Science Minister Lord Sainsbury

> *Mapping the human genome has been compared with putting a man on the moon, but I believe it is more than that. This is the outstanding achievement not only of our lifetime, but in terms of human history.*
> Michael Dexter, The Wellcome Trust

From these quotations, and many others like them, we might well think that researchers could have relaxed a bit after June 2000 because most human health and disease problems could now be sorted out really easily. After all, we had the blueprint for humankind. All we needed to do was get a bit better at understanding this set of instructions, so we could fill in a few details.

Unfortunately, these statements have proved at best premature. The reality is rather different.

We talk about DNA as if it's a template, like a mould for a car part in a factory. In the factory, molten metal or plastic gets poured into the mould thousands of times and, unless something goes wrong in the process, out pop thousands of identical car parts.

But DNA isn't really like that. It's more like a script. Think of *Romeo and Juliet*, for example. In 1936 George Cukor directed Leslie Howard and Norma Shearer in a film version. Sixty years later Baz Luhrmann directed Leonardo DiCaprio and Claire Danes in another movie version of this play. Both productions used Shakespeare's script, yet the two movies are entirely different. Identical starting points, different outcomes.

That's what happens when cells read the genetic code that's in DNA. The same script can result in different productions. The implications of this for human health are very wide-ranging, as we will see from the case studies we are going to look at in a moment. In all these case studies it's really important to remember that nothing happened to the DNA blueprint of the people in these case studies. Their DNA didn't change (mutate), and yet their life histories altered irrevocably in response to their environments.

Audrey Hepburn was one of the 20th century's greatest movie stars. Stylish, elegant and with a delicately lovely, almost fragile bone structure, her role as Holly Golightly in *Breakfast at Tiffany's* has made her an icon, even to those who have never seen the movie. It's startling to think that this wonderful beauty was created by terrible hardship. Audrey Hepburn was a survivor of an event in the Second World War known as the Dutch Hunger Winter. This ended when she was sixteen years old but the after-effects of this period, including poor physical health, stayed with her for the rest of her life.

The Dutch Hunger Winter lasted from the start of November 1944 to the late spring of 1945. This was a bitterly cold period in Western Europe, creating further hardship in a continent that had been devastated by four years of brutal war. Nowhere was this worse than in the Western Netherlands, which at this stage was still under German control. A German blockade resulted in a catastrophic drop in the availability of food to the Dutch population. At one point the population was trying to survive on only about 30 per cent of the normal daily calorie intake. People ate grass and tulip bulbs, and burned every scrap of furniture they could get their hands on, in a desperate effort to stay alive. Over 20,000 people had died by the time food supplies were restored in May 1945.

The dreadful privations of this time also created a remarkable scientific study population. The Dutch survivors were a well-defined group of individuals all of whom suffered just one period of malnutrition, all of them at exactly the same time. Because of the excellent healthcare infrastructure and record-keeping in the Netherlands, epidemiologists have been able to follow the long-term effects of the famine. Their findings were completely unexpected.

One of the first aspects they studied was the effect of the famine on the birth weights of children who had been in the womb during that terrible period. If a mother was well-fed around the time of conception and malnourished only for the last few months of the pregnancy, her baby was likely to be born small. If, on the other hand, the mother suffered malnutrition for the first three months of the pregnancy only (because the baby was conceived towards the end of this terrible episode), but then was well-fed, she was likely to have a baby with a normal body weight. The foetus 'caught up' in body weight.

That all seems quite straightforward, as we are all used to the idea that foetuses do most of their growing in the last few months of pregnancy. But epidemiologists were able to study these groups of babies for decades and what they found was really surprising. The babies who were born small stayed small all their lives, with lower obesity rates than the general population. For forty or more years, these people had access to as much food as they wanted, and yet their bodies never got over the early period of malnutrition. Why not? How did these early life experiences affect these individuals for decades? Why weren't these people able to go back to normal, once their environment reverted to how it should be?

Even more unexpectedly, the children whose mothers had been malnourished only early in pregnancy, had higher obesity rates than normal. Recent reports have shown a greater incidence of other health problems as well, including certain tests of mental activity. Even though these individuals had seemed perfectly healthy at birth, something had happened to their development in the womb that affected them for decades after. And it wasn't just the fact that something had happened that mattered, it was *when* it happened. Events that take place in the first three months of development, a stage when the foetus is really very small, can affect an individual for the rest of their life.

Even more extraordinarily, some of these effects seem to be present in the children of this group, i.e. in the grandchildren of the women who were malnourished during the first three months of their pregnancy. So something that happened in one pregnant population affected their children's children. This raised the really puzzling question of how these effects were passed on to subsequent generations.

Let's consider a different human story. Schizophrenia is a dreadful mental illness which, if untreated, can completely overwhelm and disable an affected person. Patients may present with a range of symptoms including delusions, hallucinations and enormous difficulties focusing mentally. People with schizophrenia may become completely incapable of distinguishing between the 'real world' and their own hallucinatory and delusional realm. Normal cognitive, emotional and societal responses are lost. There is a terrible misconception that people with schizophrenia are likely to be violent and dangerous. For the majority of patients this isn't the case at all, and the people most likely to suffer harm because of this illness are the patients themselves. Individuals with schizophrenia are fifty times more likely to attempt suicide than healthy individuals[2].

Schizophrenia is a tragically common condition. It affects between 0.5 per cent and 1 per cent of the population in most countries and cultures, which means that there may be over fifty million people alive today who are suffering from this condition. Scientists have known for some time that genetics plays a strong role in determining if a person will develop this illness. We know this because if one of a pair of identical twins has schizophrenia, there is a 50 per cent chance that their twin will also have the condition. This is much higher than the 1 per cent risk in the general population.

Identical twins have exactly the same genetic code as each other. They share the same womb and usually they are brought up in very similar environments. When we consider this, it doesn't seem surprising that if one of the twins develops schizophrenia, the chance that his or her twin will also develop the illness is very high. In fact, we have to start wondering why it isn't higher. Why isn't the figure 100 per cent? How is it that two apparently identical individuals can become so very different? An individual has a devastating mental illness but will their identical twin suffer from it too? Flip a coin – heads they win, tails they lose. Variations in the environment are unlikely to account for this, and even if they did, how would these environmental effects have such profoundly different impacts on two genetically identical people?

Here's a third case study. A small child, less than three years old, is abused and neglected by his or her parents. Eventually, the state intervenes and the child is taken away from the biological parents and placed with foster or adoptive parents. These new carers love and cherish the child, doing everything they can to create a secure home, full of affection. The child stays with these new parents throughout the rest of its childhood and adolescence, and into young adulthood.

Sometimes everything works out well for this person. They grow up into a happy, stable individual indistinguishable from all their peers who had normal, non-abusive childhoods. But often, tragically, it doesn't work out this way. Children who have suffered from abuse or neglect in their early years grow up with a substantially higher risk of adult mental health problems than the general population. All too often the child grows up into an adult at high risk of depression, self-harm, drug abuse and suicide.

Once again, we have to ask ourselves why. Why is it so difficult to override the effects of early childhood exposure to neglect or abuse? Why should something that happened early in life have effects on mental health that may still be obvious decades later? In some cases, the adult may have absolutely no recollection of the traumatic events, and yet they may suffer the consequences mentally and emotionally for the rest of their lives.

These three case studies seem very different on the surface. The first is mainly about nutrition, especially of the unborn child. The second is about the differences that arise between genetically identical individuals. The third is about long-term psychological damage as a result of childhood abuse.

But these stories are linked at a very fundamental biological level. They are all examples of epigenetics. Epigenetics is the new discipline that is revolutionising biology. Whenever two genetically identical individuals are non-identical in some way we can measure, this is called epigenetics. When a change in environment has biological consequences that last long after the event itself has vanished into distant memory, we are seeing an epigenetic effect in action.

Epigenetic phenomena can be seen all around us, every day. Scientists have identified many examples of epigenetics, just like the ones described above, for many years. When scientists talk about epigenetics they are referring to all the cases where the genetic code alone isn't enough to describe what's happening – there must be something else going on as well.

This is one of the ways that epigenetics is described scientifically, where things which are genetically identical can actually appear quite different to one another. But there has to be a mechanism that brings out this mismatch between the genetic script and the final outcome. These epigenetic effects must be caused by some sort of physical change, some alterations in the vast array of molecules that make up the cells of every living organism. This leads us to the other way of viewing epigenetics – the molecular description. In this model, epigenetics can be defined as the set of modifications to our genetic material that change the ways genes are switched on or off, but which don't alter the genes themselves.

Although it may seem confusing that the word 'epigenetics' can have two different meanings, it's just because we are describing the same event at two different levels. It's a bit like looking at the pictures in old newspapers with a magnifying glass, and seeing that they are made up of dots. If we didn't have a magnifying glass we might have thought that each picture was just made in one solid piece and we'd probably never have been able to work out how so many new images could be created each day. On the other hand, if all we ever did was look through the magnifying glass, all we would see would be dots, and we'd never see the incredible image that they formed together and which we'd see if we could only step back and look at the big picture.

The revolution that has happened very recently in biology is that for the first time we are actually starting to understand how amazing epigenetic phenomena are caused. We're no longer just seeing the large image, we can now also analyse the individual dots that created it. Crucially, this means that we are finally starting to unravel the missing link between nature and nurture; how our environment talks to us and alters us, sometimes forever.

The 'epi' in epigenetics is derived from Greek and means at, on, to, upon, over or beside. The DNA in our cells is not some pure, unadulterated molecule. Small chemical groups can be added at specific regions of DNA. Our DNA is also smothered in special proteins. These proteins can themselves be covered with additional small chemicals. None of these molecular amendments changes the underlying genetic code. But adding these chemical groups to the DNA, or to the associated proteins, or removing them, changes the expression of nearby genes. These changes in gene expression alter the functions of cells, and the very nature of the cells themselves. Sometimes, if these patterns of chemical modifications are put on or taken off at a critical period in development, the pattern can be set for the rest of our lives, even if we live to be over a hundred years of age.

There's no debate that the DNA blueprint is a starting point. A very important starting point and absolutely necessary, without a doubt. But it isn't a sufficient explanation for all the sometimes wonderful, sometimes awful, complexity of life. If the DNA sequence was all that mattered, identical twins would always be absolutely identical in every way. Babies born to malnourished mothers would gain weight as easily as other babies who had a healthier start in life. And as we shall see in Chapter 1, we would all look like big amorphous blobs, because all the cells in our bodies would be completely identical.

Huge areas of biology are influenced by epigenetic mechanisms, and the revolution in our thinking is spreading further and further into unexpected frontiers of life on our planet. Some of the other examples we'll meet in this book include why we can't make a baby from two sperm or two eggs, but have to have one of each. What makes cloning possible? Why is cloning so difficult? Why do some plants need a period of cold before they can flower? Since queen bees and worker bees are genetically identical, why are they completely different in form and function? Why are all tortoiseshell cats female? Why is it that humans contain trillions of cells in hundreds of complex organs, and microscopic worms contain about a thousand cells and only rudimentary organs, but we and the worm have the same number of genes?

Scientists in both the academic and commercial sectors are also waking up to the enormous impact that epigenetics has on human health. It's implicated in diseases from schizophrenia to rheumatoid arthritis, and from cancer to chronic pain. There are already two types of drugs that successfully treat certain cancers by interfering with epigenetic processes. Pharmaceutical companies are spending hundreds of millions of dollars in a race to develop the next generation of epigenetic drugs to treat some of the most serious illnesses afflicting the industrialised world. Epigenetic therapies are the new frontiers of drug discovery.

In biology, Darwin and Mendel came to define the 19th century as the era of evolution and genetics; Watson and Crick defined the 20th century as the era of DNA, and the functional understanding of how genetics and evolution interact. But in the 21st century it is the new scientific discipline of epigenetics that is unravelling so much of what we took as dogma and rebuilding it in an infinitely more varied, more complex and even more beautiful fashion.

The world of epigenetics is a fascinating one. It's filled with remarkable subtlety and complexity, and in Chapters 3 and 4 we'll delve deeper into the molecular biology of what's happening to our genes when they become epigenetically modified. But like so many of the truly revolutionary concepts in biology, epigenetics has at its basis some issues that are so simple they seem completely self-evident as soon as they are pointed out. Chapter 1 is the single most important example of such an issue. It's the investigation which started the epigenetics revolution.

Notes on nomenclature

There is an international convention on the way that the names of genes and proteins are written, which we adhere to in this book.

Gene names and symbols are written in *italics*. The proteins encoded by the genes are written in plain text.

The symbols for human genes and proteins are written in upper case. For other species, such as mice, the symbols are usually written with only the first letter capitalised.

This is summarised for a hypothetical gene in the following table.

	Human	Other e.g. mouse
Gene name	*SO DAMNED COMPLICATED*	*So Damned Complicated*
Gene symbol	*SDC*	*Sdc*
Protein name	SO DAMNED COMPLICATED	So Damned Complicated
Protein symbol	SDC	Sdc

Like all rules, however, there are a few quirks in this system and while these conventions apply in general we will encounter some exceptions in this book.

Chapter 1

An Ugly Toad and an Elegant Man

Like the toad, ugly and venomous,
Wears yet a precious jewel in his head
William Shakespeare

Humans are composed of about 50 to 70 trillion cells. That's right, 50,000,000,000,000 cells. The estimate is a bit vague but that's hardly surprising. Imagine we somehow could break a person down into all their individual cells and then count those cells, at a rate of one cell every second. Even at the lower estimate it would take us about a million and a half years, and that's without stopping for coffee or losing count at any stage. These cells form a huge range of tissues, all highly specialised and completely different from one another. Unless something has gone very seriously wrong, kidneys don't start growing out of the top of our heads and there are no teeth in our eyeballs. This seems very obvious – but why don't they? It's actually quite odd, when we remember that every cell in our body was derived from the division of just one starter cell. This single cell is called the zygote. A zygote forms when one sperm merges with one egg. This zygote splits in two; those two cells divide again and so on, to create the miraculous piece of work which is a full human body. As they divide the cells become increasingly different from one another and form specialised cell types. This process is known as differentiation. It's a vital one in the formation of any multicellular organism.

If we look at bacteria down a microscope then pretty much all the bacteria of a single species look identical. Look at certain human cells in the same way – say, a food-absorbing cell from the small intestine and a neuron from the brain – and we would be hard pressed to say that they were even from the same planet. But so what? Well, the big 'what' is that these cells started out with exactly the same genetic material as one another. And we do mean exactly – this has to be the case, because they came from just one starter cell, that zygote. So the cells have become completely different even though they came from one cell with just one blueprint.

One explanation for this is that the cells are using the same information in different ways and that's certainly true. But it's not necessarily a statement that takes us much further forwards. In a 1960 adaptation of H. G. Wells's *The Time Machine*, starring Rod Taylor as the time-travelling scientist, there's a scene where he shows his time machine to some learned colleagues (all male, naturally) and one asks for an explanation of how the machine works. Our hero then describes how the occupant of the machine will travel through time by the following mechanism:

> *In front of him is the lever that controls movement. Forward pressure sends the machine into the future. Backward pressure, into the past. And the harder the pressure, the faster the machine travels.*

Everyone nods sagely at this explanation. The only problem is that this isn't an explanation, it's just a description. And that's also true of that statement about cells using the same information in different ways – it doesn't really tell us anything, it just re-states what we already knew in a different way.

What's much more interesting is the exploration of *how* cells use the same genetic information in different ways. Perhaps even more important is how the cells remember and keep on doing it. Cells in our bone marrow keep on producing blood cells, cells in our liver keep on producing liver cells. Why does this happen?

One possible and very attractive explanation is that as cells become more specialised they rearrange their genetic material, possibly losing genes they don't require. The liver is a vital and extremely complicated organ. The website of the British Liver Trust[1] states that the liver performs over 500 functions, including processing the food that has been digested by our intestines, neutralising toxins and creating enzymes that carry out all sorts of tasks in our bodies. But one thing the liver simply never does is transport oxygen around the body. That job is carried out by our red blood cells, which are stuffed full of a particular protein, haemoglobin. Haemoglobin binds oxygen in tissues where there's lots available, like our lungs, and then releases it when the red blood cell reaches a tissue that needs this essential chemical, such as the tiny blood vessels in the tips of our toes. The liver is never going to carry out this function, so perhaps it just gets rid of the haemoglobin gene, which it simply never uses.

It's a perfectly reasonable suggestion – cells could simply lose genetic material they aren't going to use. As they differentiate, cells could jettison hundreds of genes they no longer need. There could of course be a slightly less drastic variation on this – maybe the cells shut down genes they aren't using. And maybe they do this so effectively that these genes can never ever be switched on again in that cell, i.e. the genes are irreversibly inactivated. The key experiments that examined these eminently reasonable hypotheses – loss of genes, or irreversible inactivation – involved an ugly toad and an elegant man.

Turning back the biological clock

The work has its origins in experiments performed many decades ago in England by John Gurdon, first in Oxford and subsequently Cambridge. Now Professor Sir John Gurdon, he still works in a lab in Cambridge, albeit these days in a gleaming modern building that has been named after him. He's an engaging, unassuming and striking man who, 40 years on from his ground-breaking work, continues to publish research in a field that he essentially founded.

John Gurdon cuts an instantly recognisable figure around Cambridge. Now in his seventies, he is tall, thin and has a wonderful head of swept back blonde hair. He looks like the quintessential older English gentleman of American movies, and fittingly he went to school at Eton. There is a lovely story that John Gurdon still treasures a school report from his biology teacher at that institution which says, 'I believe Gurdon has ideas about becoming a scientist. In present showing, this is quite ridiculous.'[2] The teacher's comments were based on his pupil's dislike of mindless rote learning of unconnected facts. But as we shall see, for a scientist as wonderful as John Gurdon, memory is much less important than imagination.

In 1937 the Hungarian biochemist Albert Szent-Gyorgyi won the Nobel Prize for Physiology or Medicine, his achievements including the discovery of vitamin C. In a phrase that has various subtly different translations but one consistent interpretation he defined discovery as, 'To see what everyone else has seen but to think what nobody else has thought'[3]. It is probably the best description ever written of what truly great scientists do. And John Gurdon is truly a great scientist, and may well follow in Szent-Gyorgyi's Nobel footsteps. In 2009 he was a co-recipient of the Lasker Prize, which is to the Nobel what the Golden Globes are so often to the Oscars. John Gurdon's work is so wonderful that when it is first described it seems so obvious, that anyone could have done it. The questions he asked, and the ways in which he answered them, have that scientifically beautiful feature of being so elegant that they seem entirely self-evident.

John Gurdon used non-fertilised toad eggs in his work. Any of us who has ever kept a tank full of frogspawn and watched this jelly-like mass develop into tadpoles and finally tiny frogs, has been working, whether we thought about it in these terms or not, with fertilised eggs, i.e. ones into which sperm have entered and created a new complete nucleus. The eggs John Gurdon worked on were a little like these, but hadn't been exposed to sperm.

There were good reasons why he chose to use toad eggs in his experiments. The eggs of amphibians are generally very big, are laid in large numbers outside the body and are see-through. All these features make amphibians a very handy experimental species in developmental biology, as the eggs are technically relatively easy to handle. Certainly a lot better than a human egg, which is hard to obtain, very fragile to handle, is not transparent and is so small that we need a microscope just to see it.

John Gurdon worked on the African clawed toad (*Xenopus laevis*, to give it its official title), one of those John Malkovich ugly-handsome animals, and investigated what happens to cells as they develop and differentiate and age. He wanted to see if a tissue cell from an adult toad still contained all the genetic material it had started with, or if it had lost or irreversibly inactivated some as the cell became more specialised. The way he did this was to take a nucleus from the cell of an adult toad and insert it into an unfertilised egg that had had its own nucleus removed. This technique is called somatic cell nuclear transfer (SCNT), and will come up over and over again. 'Somatic' comes from the Greek word for 'body'.

After he'd performed the SCNT, John Gurdon kept the eggs in a suitable environment (much like a child with a tank of frogspawn) and waited to see if any of these cultured eggs hatched into little toad tadpoles.

The experiments were designed to test the following hypothesis: 'As cells become more specialised (differentiated) they undergo an *irreversible* loss/inactivation of genetic material.' There were two possible outcomes to these experiments:

Either
The hypothesis was correct and the 'adult' nucleus has lost some of the original blueprint for creating a new individual. Under these circumstances an adult nucleus will never be able to replace the nucleus in an egg and so will never generate a new healthy toad, with all its varied and differentiated tissues.

Or

The hypothesis was wrong, and new toads can be created by removing the nucleus from an egg and replacing it with one from adult tissues.

Other researchers had started to look at this before John Gurdon decided to tackle the problem – two scientists called Briggs and King using a different amphibian, the frog *Rana pipiens*. In 1952 they transplanted the nuclei from cells at a very early stage of development into an egg lacking its own original nucleus and they obtained viable frogs. This demonstrated that it was technically possible to transfer a nucleus from another cell into an 'empty' egg without killing the cell. However, Briggs and King then published a second paper using the same system but transferring a nucleus from a more developed cell type and this time they couldn't create any frogs. The difference in the cells used for the nuclei in the two papers seems astonishingly minor – just one day older and no froglets. This supported the hypothesis that some sort of irreversible inactivation event had taken place as the cells differentiated. A lesser man than John Gurdon might have been put off by this. Instead he spent over a decade working on the problem.

The design of the experiments was critical. Imagine we have started reading detective stories by Agatha Christie. After we've read our first three we develop the following hypothesis: 'The killer in an Agatha Christie novel is always the doctor.' We read three more and the doctor is indeed the murderer in each. Have we proved our hypothesis? No. There's always going to be the thought that maybe we should read just one more to be sure. And what if some are out of print, or unobtainable? No matter how many we read, we may never be entirely sure that we've read the entire collection. But that's the joy of *disproving* hypotheses. All we need is one instance in which Poirot or Miss Marple reveal that the doctor was a man of perfect probity and the killer was actually the vicar, and our hypothesis is shot to pieces. And that is how the best scientific experiments are designed – to disprove, not to prove an idea.

And that was the genius of John Gurdon's work. When he performed his experiments what he was attempting was exceptionally challenging with the technology of the time. If he failed to generate toads from the adult nuclei this could simply mean his technique had something wrong with it. No matter how many times he did the experiment without getting any toads, this wouldn't actually prove the hypothesis. But if he *did* generate live toads from eggs where the original nucleus had been replaced by the adult nucleus he would have *disproved* the hypothesis. He would have demonstrated beyond doubt that when cells differentiate, their genetic material isn't irreversibly lost or changed. The beauty of this approach is that just one such toad would topple the entire theory – and topple it he did.

John Gurdon is incredibly generous in his acknowledgement of the collegiate nature of scientific research, and the benefits he obtained from being in dynamic laboratories and universities. He was lucky to start his work in a well set-up laboratory which had a new piece of equipment which produced ultraviolet light. This enabled him to kill off the original nuclei of the recipient eggs without causing too much damage, and also 'softened up' the cell so that he could use tiny glass hypodermic needles to inject donor nuclei. Other workers in the lab had, in some unrelated research, developed a strain of toads which had a mutation with an easily detectable, but non-damaging effect. Like almost all mutations this was carried in the nucleus, not the cytoplasm. The cytoplasm is the thick liquid inside cells, in which the nucleus sits. So John Gurdon used eggs from one strain and donor nuclei from the mutated strain. This way he would be able to show unequivocally that any resulting toads had been coded for by the donor nuclei, and weren't just the result of experimental error, as could happen if a few recipient nuclei had been left over after treatment.

John Gurdon spent around fifteen years, starting in the late 1950s, demonstrating that in fact nuclei from specialised cells *are* able to create whole animals if placed in the right environment i.e. an unfertilised egg[4]. The more differentiated/specialised the donor cell was, the less successful the process in terms of numbers of animals, but that's the beauty of disproving a hypothesis – we might need a lot of toad eggs to start with but we don't need to end up with many live toads to make our case. Just one non-murderous doctor will do it, remember?

So John Gurdon showed us that although there is something in cells that can keep specific genes turned on or switched off in different cell types, whatever this something is, it can't be loss or permanent inactivation of genetic material, because if he put an adult nucleus into the right environment – in this case an 'empty' unfertilised egg – it forgot all about this memory of which cell type it came from. It went back to being a naive nucleus from an embryo and started the whole developmental process again.

Epigenetics is the 'something' in these cells. The epigenetic system controls how the genes in DNA are used, in some cases for hundreds of cell division cycles, and the effects are inherited from when cells divide. Epigenetic modifications to the essential blueprint exist over and above the genetic code, on top of it, and program cells for decades. But under the right circumstances, this layer of epigenetic information can be removed to reveal the same shiny DNA sequence that was always there. That's what happened when John Gurdon placed the nuclei from fully differentiated cells into the unfertilised egg cells.

Did John Gurdon know what this process was when he generated his new baby toads? No. Does that make his achievement any less magnificent? Not at all. Darwin knew nothing about genes when he developed the theory of evolution through natural selection. Mendel knew nothing about DNA when, in an Austrian monastery garden, he developed his idea of inherited factors that are transmitted 'true' from generation to generation of peas. It doesn't matter. They saw what nobody else had seen and suddenly we all had a new way of viewing the world.

The epigenetic landscape

Oddly enough, there was a conceptual framework that was in existence when John Gurdon performed his work. Go to any conference with the word 'epigenetics' in the title and at some point one of the speakers will refer to something called 'Waddington's epigenetic landscape'. They will show the grainy image seen in Figure 1.1.

Conrad Waddington was a hugely influential British polymath. He was born in 1903 in India but was sent back to England to go to school. He studied at Cambridge University but spent most of his career at the University of Edinburgh. His academic interests ranged from developmental biology to the visual arts to philosophy, and the cross-fertilisation between these areas is evident in the new ways of thinking that he pioneered.

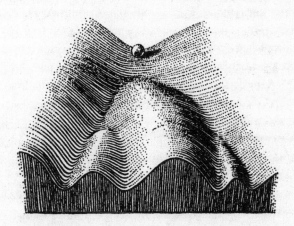

Figure 1.1   The image created by Conrad Waddington to represent the epigenetic landscape. The position of the ball represents different cell fates.

Waddington presented his metaphorical epigenetic landscape in 1957 to exemplify concepts of developmental biology[5]. The landscape merits quite a bit of discussion. As you can see, there is a ball at the top of a hill. As the ball rolls down the hill, it can roll into one of several troughs towards the bottom of the hill. Visually this immediately suggests various things to us, because we have all at some point in our childhood rolled balls down hills, or stairs, or something.

What do we immediately understand when we see the image of Waddington's landscape? We know that once a ball has reached the bottom it is likely to stay there unless we do something to it. We know that to get the ball back up to the top will be harder than rolling it down the hill in the first place. We also know that to roll the ball out of one trough and into another will be hard. It might even be easier to roll it part or all of the way back up and then direct it into a new trough, than to try and roll it directly from one trough to another. This is especially true if the two troughs we're interested in are separated by more than one hillock.

This image is incredibly powerful in helping to visualise what might be happening during cellular development. The ball at the top of the hill is the zygote, the single cell that results from the fusion of one egg and one sperm. As the various cells of the body begin to differentiate (become more specialised), each cell is like a ball that has rolled further down the hill and headed into one of the troughs. Once it has gone as far as it can go, it's going to stay there. Unless something extraordinarily dramatic happens, that cell is never going to turn into another cell type (jump across to another trough). Nor is it going to move back up to the top of the hill and then roll down again to give rise to all sorts of different cell types.

Like the time traveller's levers, Waddington's landscape at first just seems like another description. But it's more than that, it's a model that helps us to develop ways of thinking. Just like so many of the scientists in this chapter, Waddington didn't know the details of the mechanisms but that didn't really matter. He gave us a way of thinking about a problem that was useful.

John Gurdon's experiments had shown that sometimes, if he pushed hard enough, he could move a cell from the very bottom of a trough at the bottom of the hill, right the way back up to the top. From there it can roll down and become any other cell type once more. And every toad that John Gurdon and his team created taught us two other important things. The first is that cloning – the recreation of an animal from the cells of an adult – is possible, because that's what he had achieved. The second thing it taught us is that cloning is really difficult, because he had to perform hundreds of SCNTs for every toad that he managed to generate.

That's why there was such a furore in 1996 when Keith Campbell and Ian Wilmut at the Roslin Institute created the first mammalian clone, Dolly the sheep[6]. Like John Gurdon, they used SCNT. In the case of Dolly, the scientists transferred the nucleus from a cell in the mammary gland of an adult ewe into an unfertilised sheep egg from which they had removed the original nucleus. Then they transplanted this into the uterus of a recipient ewe. Pioneers of cloning were nothing if not obsessively persistent. Campbell and Wilmut performed nearly 300 nuclear transfers before they obtained that one iconic animal, which now revolves in a glass case in the Royal Scottish Museum in Edinburgh. Even today, when all sorts of animals have been cloned, from racehorses to prize cattle and even pet dogs and cats, the process is incredibly inefficient. Two questions have remained remarkably pertinent since Dolly tottered on her soon to be prematurely arthritic legs into the pages of history. The first is why is cloning animals so inefficient? The second is why are the animals so often less healthy than 'natural' offspring? The answer in both cases is epigenetics, and the molecular explanations will become apparent as we move through our exploration of the field. But before we do, we're going to take our cue from H. G. Wells's time traveller and fast-forward over thirty years from John Gurdon in Cambridge to a laboratory in Japan, where an equally obsessive scientist has found a completely new way of cloning animals from adult cells.

Chapter 2

How We Learned
to Roll Uphill

*Any intelligent fool can make things bigger and more
complex … It takes a touch of genius and a lot of
courage to move in the opposite direction.*
Albert Einstein

Let's move on about 40 years from John Gurdon's work, and a
decade on from Dolly. There is so much coverage in the press
about cloned mammals that we might think this procedure has
become routine and easy. The reality is that it is still highly time-
consuming and laborious to create clones by nuclear transfer,
and consequently it's generally a very costly process. Much of the
problem lies in the fact that the process relies on manually transferring somatic nuclei into eggs. Unlike the amphibians that John
Gurdon worked on, there's the additional problem that mammals
don't produce very many eggs at once. Mammalian eggs also have
to be extracted carefully from the body, they aren't just ejected into
a tank like toad eggs. Mammalian eggs have to be cultured incredibly delicately to keep them healthy and alive. Researchers need
to remove the nucleus manually from an egg, inject in a nucleus
from an adult cell (without damaging anything), then keep culturing the cells really, really carefully until they can be implanted
into the uterus of another female. This is incredibly intensive and
painstaking work and we can only do it one cell at a time.

For many years, scientists had a dream of how they would
carry out cloning in an ideal world. They would take really accessible cells from the adult mammal they wanted to clone. A small

sample of cells scraped from the skin would be a pleasantly easy option. Then they would treat these cells in the laboratory, adding specific genes, or proteins, or chemicals. This treatment would change the way the nuclei of these cells behaved. Instead of acting like the nucleus of a skin cell, they would act the same way as nuclei from newly fertilised eggs. The treatment would therefore have the same ultimate effect as transferring the nuclei from adult cells into fertilised eggs, from which their own nuclei had been removed. The beauty of such a hypothetical scheme is that we'd have bypassed most of the really difficult and time-consuming steps that require such a high level of technical skill in manipulating tiny cells. This would make it an easily accessible technique and one that could be carried out on lots of cells simultaneously, rather than just one nuclear transfer at a time.

Okay, we'd still have to find a way of putting them into a surrogate mother, but we only have to go down the surrogate mother route if we want to generate a complete individual. Sometimes this is exactly what we want – to re-create a prize bull or prize stallion, for example, but this is not what most sane people want to do with humans. Indeed cloning humans (reproductive cloning) is banned in pretty much every country which has the scientists and the infrastructure to undertake such a task. But actually for most purposes we don't need to go as far as this stage for cloning to be useful for humans. What we need are cells that have the potential to turn into lots of other cell types. These are the cells that are known as stem cells, and they are metaphorically near the top of Waddington's epigenetic landscape. The reason we need such cells lies in the nature of the diseases that are major problems in the developed world.

In the rich parts of our planet the diseases that kill most of us are chronic. They take a long time to develop and often they take a long time to kill us when they do. Take heart disease, for example – if someone survives the initial heart attack they don't necessarily ever go back to having a totally healthy heart again.

During the attack some of the heart muscle cells (cardiomyocytes) may become starved of oxygen and die. We might imagine this would be no problem, as surely the heart can create replacement cells? After all, if we donate blood, our bone marrow can make more red blood cells. Similarly, we have to do an awful lot of damage to the liver before it stops being able to regenerate and repair itself. But the heart is different. Cardiomyocytes are referred to as 'terminally differentiated' – they have gone right to the bottom of Waddington's hill and are stuck in a particular trough. Unlike bone marrow or liver, the heart doesn't have an accessible reservoir of less specialised cells (cardiac stem cells) that could turn into new cardiomyocytes. So, the long-term problem that follows a heart attack is that our bodies can't make new cardiac muscle cells. The body does the only thing it can and replaces the dead cardiomyocytes with connective tissue, and the heart never beats in quite the same way it did before.

Similar things happen in so many diseases – the insulin-secreting cells that are lost when teenagers develop type 1 diabetes, the brain cells that are lost in Alzheimer's disease, the cartilage producing cells that disappear during osteoarthritis – the list goes on and on. It would be great if we could replace these with new cells, identical to our own. This way we wouldn't have to deal with all the rejection issues that make organ transplants such a challenge, or with the lack of availability of donors. Using stem cells in this way is referred to as therapeutic cloning; creating cells identical to a specific individual in order to treat a disease.

For over 40 years we've known that in theory this could be possible. John Gurdon's work and all that followed after him showed that adult cells contain the blueprints for all the cells of the body if we can only find the correct way of accessing them. John Gurdon had taken nuclei from adult toads, put them into toad eggs and been able to push those nuclei all the way back up Waddington's landscape and create new animals. The adult nuclei had been – and this word is critical – reprogrammed. Ian Wilmut and Keith Campbell had done pretty much the same thing with sheep. The important common feature to recognise here is that in each case the reprogramming only worked when the adult nucleus was placed inside an unfertilised egg. It was the egg that was really important. We can't clone an animal by taking an adult nucleus and putting it into some other cell type.

Why not?

We need a little cell biology here. The nucleus contains the vast majority of the DNA/genes that encode us – our blueprint. There's a miniscule fraction of DNA that isn't in the nucleus, it's in tiny structures called mitochondria, but we don't need to worry about that here. When we're first taught about cells in school it's almost as if the nucleus is all powerful and the rest of the cell – the cytoplasm – is a bag of liquid that doesn't really do much. Nothing could be further from the truth, and this is especially the case for the egg, because the toads and Dolly have taught us that the cytoplasm of the egg is absolutely key. Something, or some things, in that egg cytoplasm actively reprogrammed the adult nucleus that the experimenters injected into it. These unknown factors moved a nucleus from the bottom of one of Waddington's troughs right back to the top of the landscape.

Nobody really understood how the cytoplasm of eggs could convert adult nuclei into ones like zygotes. There was pretty much an assumption that whatever it was must be incredibly complicated and difficult to unravel. Often in science really big questions have smaller, more manageable questions inside them. So a number of labs tackled a conceptually simpler, but technically still hugely challenging issue.

Endless potential

Remember that ball at the top of Waddington's landscape. In cellular terms it's the zygote and it's referred to as totipotent, that is, it has the potential to form every cell in the body, including the placenta. Of course, zygotes by definition are rather limited in number and most scientists working in very early development use cells from a bit later, the famous embryonic stem (ES) cells. These are created as a result of normal developmental pathways. The zygote divides a few times to create a bundle of cells called the blastocyst. Although the blastocyst typically has less than 150 cells it's already an early embryo with two distinct compartments. There's an outer layer called the trophectoderm, which will eventually form the placenta and other extra-embryonic tissues, and an inner cell mass (ICM).

Figure 2.1 shows what the blastocyst looks like. The drawing is in two dimensions but in reality the blastocyst is a three-dimensional structure, so the actual shape is that of a tennis ball that's had a golf ball glued inside it.

The cells of the ICM can be grown in the lab in culture dishes. They're fiddly to maintain and require specialised culture conditions and careful handling, but do it right and they reward us

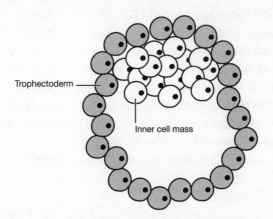

Figure 2.1 A diagram of the mammalian blastocyst. The cells of the trophectoderm will give rise to the placenta. During normal development, the cells of the Inner Cell Mass (ICM) will give rise to the tissues of the embryo. Under laboratory conditions, the cells of the ICM can be grown in culture as pluripotent embryonic stem (ES) cells.

by dividing a limitless number of times and staying the same as the parent cell. These are the ES cells and as their full name suggests, they can form every cell of the embryo and ultimately of the mature animal. They aren't totipotent – they can't make placenta – so they are called pluripotent because they make pretty much anything else.

These ES cells have been invaluable for understanding what's important for keeping cells in a pluripotent state. Over the years a number of leading scientists including Azim Surani in Cambridge, Austin Smith in Edinburgh, Rudolf Jaenisch in Boston and Shinya Yamanaka in Kyoto have devoted huge amounts of time to identifying the genes and proteins expressed (switched on) in ES cells. They particularly tried to identify genes that keep the ES cells in a pluripotent state. These genes are extraordinarily important because ES cells seem to be very prone to turn into other cell types in culture if you don't keep the conditions just right. Just a small change in culture conditions, for example, and a culture dish full of one-time ES cells can differentiate into cardiomyocytes and do what heart cells do best: they beat along in time with one another. A slightly different change in conditions – altering the delicate balance of chemicals in the culture fluid, for example, can divert the ES cells away from the cardiac route and start the development of cells that give rise to the neurons in our brains.

Scientists working on ES cells identified a whole slew of genes that were important for keeping the cells pluripotent. The functions of the various genes they identified weren't necessarily identical. Some were important for self-renewal, i.e. one ES dividing to form two ES cells, whereas others were required to stop the cells from differentiating[1].

So, by the early years of the 21st century scientists had found a way of maintaining pluripotent ES cells in culture dishes and they knew quite a lot about their biology. They had also worked out how to change the culture conditions so that the ES cells would differentiate into various cell types including liver cells, heart cells, neurons etc. But how does this help with the dream we laid out earlier? Could the labs use this information to create new ways of driving cells backwards, to the top of Waddington's landscape? Would it be possible to take a fully differentiated cell and treat it in a lab so that it would become just like an ES cell, with all the potential that implies? Whilst scientists had good reason to believe this would be theoretically possible, that's a long way from actually being able to do it. But it was a wonderfully tantalising prospect for scientists interested in using stem cells to treat human diseases.

By the middle of the first decade of this century, over twenty genes had been identified that seemed to be critical to ES cells. It wasn't necessarily clear how they worked together and there was every reason to think that there was still plenty we didn't understand about the biology of ES cells. It was assumed that it would be almost inconceivably difficult to take a mature cell and essentially recreate the vastly complex intracellular conditions that are found in an ES cell.

The triumph of optimism

Sometimes the greatest scientific breakthroughs happen because someone ignores the prevailing pessimism. In this case, the optimist who decided to test what everyone else had assumed was impossible was the aforementioned Shinya Yamanaka, with his postdoctoral research associate Kazutoshi Takahashi.

Professor Yamanaka is one of the youngest luminaries in the stem cell and pluripotency field. He was born in Osaka in the early 1960s and rather unusually he has held successful academic positions in high profile institutions in both Japan and the USA. He originally trained as a clinician and became an orthopaedic surgeon. Specialists in this discipline are sometimes dismissed by other surgeons as 'the hammer and chisel brigade'. This is unfair, but it is true that orthopaedic surgical practice is about as far away from elegant molecular biology and stem cell science as it's possible to get.

Perhaps more than any of the other researchers working in the stem cell field, Professor Yamanaka had been driven by a desire to find a way of creating pluripotent cells from differentiated cells in a lab. He started this stage of his work with a list of 24 genes which were vitally important in ES cells. These were all genes called 'pluripotency genes' – they have to be switched on if ES cells are to remain pluripotent. If you use various experimental techniques to switch these genes off, the ES cells start to differentiate, just like those beating heart cells in the culture dish, and they never revert to being ES cells again. Indeed, that is partly what happens quite naturally during mammalian development, when cells differentiate and become specialised – they switch off these pluripotency genes.

Shinya Yamanaka decided to test if combinations of these genes would drive differentiated cells backwards to a more primitive developmental stage. It seemed a long shot and there was always the worry that if the results were negative – i.e. if none of the cells went 'backwards' – he wouldn't know if it was because it just wasn't possible or if he just hadn't got the experimental conditions right. This was a risk for an established scientist like Yamanaka, but it was an even bigger gamble for a relatively junior associate like Takahashi, because of the way that the scientific career ladder works.

When faced with the exposure of damaging personal love letters, the Duke of Wellington famously responded, 'Publish and be damned!' The mantra for scientists is almost the same but differs in one critical respect. For us, it's 'publish *or* be damned' – if you don't publish papers, you can't get research funding and you can't get jobs in universities. And it is rare indeed to get a paper into a good journal if the message of your years of effort boils down to, 'I tried and I tried but it didn't work.' So to take on a project with relatively little likelihood of positive results is a huge leap of faith and we have to admire Takahashi's courage, in particular.

Yamanaka and Takahashi chose their 24 genes and decided to test them in a cell type known as MEFs – mouse embryonic fibroblasts. Fibroblasts are the main cells in connective tissue and are found in all sorts of organs including skin. They're really easy to extract and they grow very easily in culture, so are a great source of cells for experiments. Because the ones known as MEFs are from embryos the hope was that they would still retain a bit of capacity to revert to very early cell types under the right conditions.

Remember how John Gurdon used donor and acceptor toad strains that had different genetically-encoded markers, so he could tell which nuclei had generated the new animals? Yamanaka did something similar. He used cells from mice which had an extra gene added. This gene is called the neomycin resistance (neo^R) gene and it does exactly what it says on the can. Neomycin is an antibiotic-type compound that normally kills mammalian cells. But if the cells have been genetically engineered to express the neo^R gene, they will survive. When Yamanaka created the mice he needed for his experiments he inserted the neo^R gene in a particular way. This meant that the neo^R gene would only get switched on if the cell it was in had become pluripotent. The cell had to be behaving like an ES cell. So if his experiments to push the fibroblasts backwards experimentally into the undifferentiated ES cell state were successful, the cells would keep growing, even when a lethal dose of the antibiotic was added. If the experiments were unsuccessful, all the cells would die.

Professor Yamanaka and Doctor Takahashi inserted the 24 genes they wanted to test into specially designed molecules called vectors. These act like Trojan horses, carrying high concentrations of the 'extra' DNA into the fibroblasts. Once in the cell, the genes were switched on and produced their specific proteins. Introducing these vectors can be done relatively easily on a large number of cells at once, using chemical treatments or electrical pulses (no fiddly micro-injections for Yamanaka, no indeed).

When Shinya Yamanaka used all 24 genes simultaneously, some of the cells survived the neomycin treatment. It was only a tiny fraction of the cells but it was an encouraging result nonetheless. It meant these cells had switched on the neo^R gene. This implied they were behaving like ES cells. But if he used the genes singly, no cells survived. Shinya Yamanaka and Kazutoshi Takahashi then added various sets of 23 genes to the cells. They used the results from these experiments to identify ten genes that were each really critical for creating the neomycin-resistant pluripotent cells. By testing various combinations from these ten genes they finally hit on the smallest number of genes that could act together to turn embryonic fibroblasts into ES-like cells.

The magic number turned out to be four. When the fibroblasts were invaded by vectors carrying genes called *Oct4*, *Sox2*, *Klf4* and *c-Myc* something quite extraordinary happened. The cells survived in neomycin, showing they had switched on the neo^R gene and were therefore like ES cells. Not only that, but the fibroblasts began to change shape to look like ES cells. Using various experimental systems, the researchers were able to turn these reprogrammed cells into the three major tissue types from which all organs of the mammalian body are formed – ectoderm, mesoderm and endoderm. Normal ES cells can also do this. Fibroblasts never can. Shinya Yamanaka then showed that he could repeat the whole process using fibroblasts from adult mice rather than embryos as his starting material. This showed that his method didn't rely on some special feature of embryonic cells, but could also be applied to cells from completely differentiated and mature organisms.

Yamanaka called the cells that he created 'induced pluripotent stem cells' and the acronym – iPS cells – is now familiar terminology to everyone working in biology. When we consider that this phrase didn't even exist five years ago, its universal recognition amongst scientists shows just how important a breakthrough this really is.

It's incredible to think that mammalian cells carry about 20,000 genes, and yet it only takes four to turn a fully differentiated cell into something that is pluripotent. With just four genes Professor Yamanaka was able to push the ball right from the bottom of one of Waddington's troughs, all the way back up to the top of the landscape.

It wasn't surprising that Shinya Yamanaka and Kazutoshi Takahashi published their findings in *Cell*, the world's most prestigious biological journal[2]. What was a bit surprising was the reaction. Everyone in 2006 knew this was huge, but they knew it was only huge if it was right. An awful lot of scientists couldn't really believe that it was. They didn't for one moment think that Professor Yamanaka and Doctor Takahashi were lying, or had done anything fraudulent. They just thought they had probably got something wrong, because really, it couldn't be that simple. It was analogous to someone searching for the Holy Grail and finding it the second place they looked, under the peas at the back of the freezer.

The obvious thing of course would be for someone to repeat Yamanaka's work and see if they could get the same results. It may seem odd to people working outside science, but there wasn't an avalanche of labs that wanted to do this. It had taken Shinya Yamanaka and Kazutoshi Takahashi two years to run their experiments, which were time-consuming and required meticulous control of all stages. Labs would also be heavily committed to their existing programmes of research and didn't necessarily want to be diverted. Additionally, the organisations that fund researchers to carry out specific programmes of work are apt to look a bit askance if a lab head suddenly abandons a programme of agreed research to do something entirely different. This would be particularly damaging if the end result was a load of negative data. Effectively, that meant that only an exceptionally well-funded lab, with the best equipment and a very self-confident head, would even think of 'wasting time' repeating someone else's experiments.

Rudolf Jaenisch from The Whitehead Institute in Cambridge, MA is a colossus in the field of creating genetically engineered animals. Originally from Germany, he has worked in the USA for almost the last 30 years. With curly grey hair and a frankly impressive moustache, he is immediately recognisable at conferences. It was perhaps unsurprising that he was the scientist who took the risk of diverting some of the work in his lab to see if Shinya Yamanaka really had achieved the seemingly impossible. After all, Rudolf Jaenisch is on record stating that, 'I have done many high risk projects through the years, but I believe that if you have an exciting idea, you must live with the chance of failure and pursue the experiment.'

At a conference in Colorado in April 2007 Professor Jaenisch stood up to give his presentation and announced that he had repeated Yamanaka's experiments. They worked. Yamanaka was right. You could make iPS cells by introducing just four genes into a differentiated cell. The effect on the audience was dramatic. The atmosphere was like one of those great moments in old movies where the jury delivers its verdict and all the hacks dash off to call the editor.

Rudolf Jaenisch was gracious – he freely conceded that he had carried out the experiments because he just knew that Yamanaka couldn't be right. The field went crazy after that. First, the really big labs involved in stem cell research started using Yamanaka's technique, refining and improving it so it worked more efficiently. Within a couple of years even labs that had never cultured a single ES cell were generating iPS cells from tissues and donors they were interested in. Papers on iPS cells are now published every week of the year. The technique has been adapted for direct conversion of human fibroblasts into human neuronal cells without having to create iPS cells first[3]. This is equivalent to rolling a ball halfway up Waddington's epigenetic landscape and then back down into a different trough.

It's hard not to wonder if it was frustrating for Shinya Yamanaka that nobody else seemed to take up his work until the

American laboratory showed that he was right. He shared the 2009 Lasker Prize with John Gurdon so maybe he's not really all that concerned. His reputation is now assured.

Follow the money

If all we read is the scientific literature, then the narrative for this story is quite inspiring and fairly straightforward. But there's another source of information, and that's the patent landscape, which typically doesn't emerge from the mist until some time after the papers in the peer-reviewed journals. Once the patent applications in this field started appearing, a somewhat more complicated tale began to unfold. It takes a while for this to happen, because patents remain confidential for the first year to eighteen months after they are submitted to the patent offices. This is to protect the interests of the inventors, as this period of grace gives them time to get on with work on confidential areas without declaring to the world what they've invented. The important thing to realise is that both Yamanaka and Jaenisch have filed patents on their research into controlling cell fate. Both of these patent applications have been granted and it is likely that cases will go to court to test who can really get protection for what. And the odd thing, given that Yamanaka *published* first, is the fact that Jaenisch filed a patent on this field before him.

How could that be? It's partly because a patent application can be quite speculative. The applicant doesn't have to have proof of every single thing that they claim. They can use the grace period to try to obtain some proof to support their assertions from the original claim. In US legal terms Shinya Yamanaka's patent dates from 13 December 2005 and covers the work described a few paragraphs ago – how to take a somatic cell and use the four factors – *Oct4*, *Sox2*, *Klf4* and *c-Myc* – to turn it into a pluripotent cell. Rudolf Jaenisch's patent potentially could have a legal first date of 26 November 2003. It contains a number of technical aspects and it makes claims around expressing a pluripotency gene in a somatic cell. One of the genes it suggests is *Oct4*. *Oct4* had been known for some time to be vital for the pluripotent state, after all, that's one of the reasons why Yamanaka had included it in his original reprogramming experiments. The legal arguments around these patents are likely to run and run.

But why did these labs, run by fabulous and highly creative scientists, file these patents in the first place? Theoretically, a patent allows the holder access to an exclusive means of doing something. However, in academic circles nobody ever tries to stop an academic scientist in another lab from running a basic science experiment. What the patent is really for is to make sure that the original inventor makes money out of their good idea, instead of other people cashing in on their inventiveness.

The most profitable patents of all in biology tend to be things that can be used to treat disease in people, or that help researchers to develop new treatments faster. And that's why there is going to be such a battle over the Jaenisch and Yamanaka patents. The courts may decide that every time someone makes iPS cells, money will have to be paid to the researchers and institutions who own the original ideas. If companies sell iPS cells that they make, and have to give a percentage of the income back to the patent holders, the potential returns could be substantial. It's worth looking at why these cells are viewed as potentially so valuable in monetary terms.

Let's take just one disease, type 1 diabetes. This typically starts in childhood when certain cells in the pancreas (the delightfully named beta cells in the Islets of Langerhans) are destroyed through processes that aren't yet clear. Once lost, these cells never grow back and as a consequence the patient is no longer able to produce the hormone insulin. Without insulin it's impossible to control blood sugar levels and the consequences of this are potentially catastrophic. Until we found ways of extracting insulin from pigs and administering it to patients, children and young adults routinely died as a result of diabetes. Even now, when we can administer insulin relatively easily (normally an artificially synthesised human form), there are a lot of drawbacks. Patients have to monitor their blood sugar levels multiple times a day and alter their insulin dose and food intake to try and stay within certain boundaries. It's hard to do this consistently over many years, especially for a teenager. How many adolescents are motivated by things that might go wrong when they are 40? Long-term type 1 diabetics are prone to a vast range of complications, including loss of vision, poor circulation that can lead to amputations, and kidney disease.

It would be great if, instead of injecting insulin every day, diabetics could just receive new beta cells. The patient could then produce their own insulin once more. The body's own internal mechanisms are usually really good at controlling blood sugar levels so most of the complications would probably be avoided. The problem is that there are no cells in the body that are able to create beta cells (they are at the bottom of one of Waddington's troughs) so we would need to use either a pancreas transplant or perhaps change some human ES cells into beta cells and put those into the patient.

There are two big problems in doing this. The first is that donor materials (either ES cells or a whole pancreas) are in short supply so there's nowhere near enough to supply all the diabetics. But even if there were enough, there's still the problem that they won't be the same as the patient's tissues. The patient's immune system will recognise them as foreign and try to reject them. The person might be able to come off insulin but would probably need to be on immuno-suppressive drugs all their life. This is not really that much of a trade-off, as these drugs have a range of pretty awful side-effects.

iPS cells suddenly create a new way forwards. Take a small scraping of skin cells from our patient, whom we shall call Freddy. Grow these cells in culture until we have enough to work with (this is pretty easy). Use the four Yamanaka factors to create a large number of iPS cells, treat these in the lab to turn them into beta cells and put them back into the patient. There will be no immune rejection because Freddy will just be receiving Freddy cells. Recently, researchers have shown they can do exactly this in mouse models of diabetes[4].

It won't be that simple of course. There are a whole range of technological hurdles to overcome, not least the fact that one of the four Yamanaka factors, c-Myc, is known to promote cancer. But in the few years since that key publication in *Cell*, substantial progress has been made in improving the technology so that it is moving ever closer to the clinic. It's possible to make human iPS cells pretty much as easily as mouse ones and you don't always need to use c-Myc[5]. There are ways of creating the cells that take away some of the other worrying safety problems as well. For example, the first methods for creating iPS cells used animal products in the cell culture stages. This is always a worry, because of fears about transmitting weird animal diseases into the human population. But researchers have now found synthetic replacements for these animal products[6]. The whole field of iPS production is getting better all the time. But we're not over the line yet.

One of the problems commercially is that we don't yet know what the regulatory authorities will demand by way of safety and supporting data before they let iPS cells be used in humans. Currently, licensing iPS cells for therapeutic use would involve two different areas of medical regulation. This is because we would be giving a patient cells (cell therapy) which had been genetically modified (gene therapy). Regulators are wary particularly because so many of the gene therapy trials that were launched with such enthusiasm in the 1980s and 1990s either had little benefit for the patient or sometimes even terrible and unforeseen consequences, including induction of lethal cancers[7]. The number of potentially costly regulatory hurdles iPS cells will have to get over before they can be given to patients is huge. We might think no investor would put any money into something so potentially risky. Yet invest they do, and that's because if researchers can get this technology right the return on the investment could be huge.

Here's just one calculation. At a conservative estimate, it costs about $500 per month in the United States to supply insulin and blood sugar monitoring equipment for a diabetic. That's $6,000 a year, so if a patient lives with diabetes for 40 years that's $240,000 over their lifetime. Then add in the costs of all the treatments that even well-managed diabetic patients will need for the complications they are likely to suffer because of their illness. It's fairly easy to see how each patient's diabetes-related lifetime healthcare costs could be at least a million dollars. And there are at least a million type 1 diabetics in the US alone. This means that at the very least, the US economy spends over a billion dollars every four years, just in treating type 1 diabetes. So even if iPS cells cost a lot to get into the clinic, they have the potential to make an enormous return on investment if they work out cheaper than the lifetime cost of current therapies.

That's just for diabetes. There are a whole host of other diseases for which iPS cells could provide an answer. Just a few examples include patients with blood clotting disorders, such as haemophilias; Parkinson's disease; osteo-arthritis and blindness caused by macular degeneration. As science and technology get better at creating artificial structures that can be implanted into our bodies, iPS cells will be used for replacing damaged blood vessels in heart disease, and regenerating tissues destroyed by cancer or its treatment.

The US Department of Defense is providing funding into iPS cells. The military always needs plenty of blood in any combat situation so that it can treat wounded personnel. Red blood cells aren't like most cells in our bodies. They have no nucleus, which means they can't divide to form new cells. This makes red blood cells a relatively safe type of iPS cell to start using clinically, as they won't stay in the body for more than a few weeks. We also don't reject these cells in the same way that we would a donor kidney, for example, because there are differences in the ways our immune systems recognise these cells. Different people can have compatible red blood cells – it's the famous ABO blood type system, plus some added complications. It's been calculated that we could take just 40 donors of specific blood types, and create a bank of iPS cells from those people that would supply all our needs[8]. Because iPS cells can keep on dividing to create more iPS cells when grown under the right conditions, we could create a never-ending bank of cells. There are well-established methods for taking immature blood stem cells and growing them under specific stimuli so that they will differentiate to form (ultimately) red blood cells. Essentially, it should be possible to create a huge bank of different types of red blood cells, so that we can always have matching blood for patients, be these from the battlefield or a traffic accident.

The generation of iPS cells has been one of those rare events in biology that have not just changed a field, but have almost reinvented it. Shinya Yamanaka is considered by most to be a dead cert to share a Nobel Prize with John Gurdon in the near future, and it would be difficult to over-estimate the technological impact of the work. But even though the achievement is extraordinary, nature already does so much more, so much faster.

When a sperm and an egg fuse, the two nuclei are reprogrammed by the cytoplasm of the egg. The sperm nucleus, in particular, very quickly loses most of the molecular memory of what it was and becomes an almost blank canvas. It's this reprogramming phenomenon that was exploited by John Gurdon, and by Ian Wilmut and Keith Campbell, when they inserted adult nuclei into the cytoplasm of eggs and created new clones.

When an egg and sperm fuse, the reprogramming process is incredibly efficient and is all over within 36 hours. When Shinya Yamanaka first created iPS cells only a miniscule number, a fraction far less than 1 per cent of the cells in the best experiment,

were reprogrammed. It literally took weeks for the first reprogrammed iPS cells to grow. A lot of progress has been made in
improving the percentage efficiency and speed of reprogramming
adult cells into iPS cells, but it still doesn't come within spitting
range of what happens during normal fertilisation. Why not?

The answer is epigenetics. Differentiated cells are epigenetically
modified in specific ways, at a molecular level. This is why skin
fibroblasts will normally always remain as skin fibroblasts and
not turn into cardiomyocytes, for example. When differentiated
cells are reprogrammed to become pluripotent cells – whether by
somatic cell nuclear transfer or by the use of the four Yamanaka
factors – the differentiation-specific epigenetic signature must be
removed so that the nucleus becomes more like that of a newly
fertilised zygote.

The cytoplasm of an egg is incredibly efficient at reversing
the epigenetic memory on our genes, acting as a giant molecular
eraser. This is what it does very rapidly when the egg and sperm
nuclei fuse to form a zygote. Artificial reprogramming to create
iPS cells is more like watching a six-year-old doing their home-
work – they are forever rubbing out the wrong bit whilst leaving
in the mis-spelt words, and then tearing a hole in the page because
they rub too vigorously. Although we are starting to get a handle
on some of the processes involved, we are a long way from recre-
ating in the lab what happens naturally.

Until now we have been talking about epigenetics at the phenomenon scale. The time has come to move into the molecules
that underlie all the remarkable events we've talked about so far,
and many more besides.

Chapter 3

Life As We Knew It

A poet can survive everything but a misprint.
Oscar Wilde

If we are going to understand epigenetics, we first need to understand a bit about genetics and genes. The basic code for pretty much all independent life on earth, from bacteria to elephants, from Japanese knotweed to humans, is DNA (deoxyribonucleic acid). The phrase 'DNA' has become an expression in its own right with increasingly vague meanings. Social commentators may refer to the DNA of a society or of a corporation, by which they mean the real core of values behind an organisation. There's even been a perfume called after it. The iconic scientific image of the mid-20th century was the atomic mushroom cloud. The double helix of DNA had similar cachet in the later part of the same century.

Science is just as prone to mood swings and fashions as any other human activity. There was a period when the prevailing orthodoxy seemed to be that the only thing that mattered was our DNA script, our genetic inheritance. Chapters 1 and 2 showed that this can't be the case, as the same script is used differently depending on its cellular context. The field is now possibly at risk of swinging a bit too far in the opposite direction, with hard-line epigeneticists almost minimizing the significance of the DNA code. The truth is, of course, somewhere in between.

In the Introduction, we described DNA as a script. In the theatre, if a script is lousy then even a wonderful director and a terrific cast won't be able to create a great production. On the other hand, we have probably all suffered through terrible productions of our favourite plays. Even if the script is perfect, the final outcome can be awful if the interpretation is poor. In the same way, genetics and epigenetics work intimately together to create the miracles that are us and every organic thing around us.

DNA is the fundamental information source in our cells, their basic blueprint. DNA itself isn't the real business end of things, in the sense that it doesn't carry out all the thousands of activities required just to keep us alive. That job is mainly performed by the proteins. It's proteins that carry oxygen around our bloodstream, that turn chips and burgers into sugars and other nutrients that can be absorbed from our guts and used to power our brains, that contract our muscles so we can turn the pages of this book. But DNA is what carries the codes for all these proteins.

If DNA is a code, then it must contain symbols that can be read. It must act like a language. This is indeed exactly what the DNA code does. It might seem odd when we think how complicated we humans are, but our DNA is a language with only four letters. These letters are known as bases, and their full names are adenine, cytosine, guanine and thymine. They are abbreviated to A, C, G and T. It's worth remembering C, cytosine, in particular, because this is the most important of all the bases in epigenetics.

One of the easiest ways to visualise DNA mentally is as a zip. It's not a perfect analogy, but it will get us started. Of course, one of the most obvious things that we know about a zip is that it is formed of two strips facing each other. This is also true of DNA. The four bases of DNA are the teeth on the zip. The bases on each side of the zip can link up to each other chemically and hold the zip together. Two bases facing each other and joined up like this are known as a base-pair. The fabric strips that the teeth are stitched on to on a zip are the DNA backbones. There are always two backbones facing each other, like the two sides of the zip, and DNA is therefore referred to as double-stranded. The two sides of the zip are basically twisted around to form a spiral structure – the famous double helix. Figure 3.1 is a stylised representation of what the DNA double helix looks like.

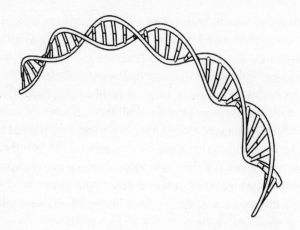

Figure 3.1   A schematic representation of DNA. The two backbones are twisted around each other to form a double helix. The helix is held together by chemical bonds between the bases in the centre of the molecule.

The analogy will only get us so far, however, and that's because the teeth of the DNA zip aren't all equivalent. If one of the teeth is an A base, it can only link up with a T base on the opposite strand. Similarly, if there is a G base on one strand, it can only link up with a C on the other one. This is known as the base-pairing principle. If an A tried to link with a C on the opposite strand it would throw the whole shape of the DNA out of kilter, a bit like a faulty tooth on a zip.

Keeping it pure

The base-pairing principle is incredibly important in terms of DNA function. During development, and even during a lot of adult life, the cells of our bodies divide. They do this so that organs can get bigger as a baby matures, for example. They also grow to replace cells that die off quite naturally. An example of this is the production by the bone marrow of white blood cells, produced to replace those that are lost in our bodies' constant battles with infectious micro-organisms. The majority of cell types reproduce by first copying their entire DNA, and then dividing it equally between two daughter cells. This DNA replication is essential. Without it, daughter cells could end up with no DNA, which in most cases would render them completely useless, like a computer that's lost its operating software.

It's the copying of DNA before each cell division that shows why the base-pairing principle is so important. Hundreds of scientists have spent their entire careers working out the details of how DNA gets faithfully copied. Here's the gist of it. The two strands of DNA are pulled apart and then the huge number of proteins involved in the copying (known as the replication complex) get to work.

Figure 3.2 shows in principle what happens. The replication complex moves along each single strand of DNA, and builds up a new strand facing it. The complex recognises a specific base – base C for example – and always puts a G in the opposite position on the strand that it's building. That's why the base-pairing principle is so important. Because C has to pair up with G, and A has to pair up with T, the cells can use the existing DNA as a template to make the new strands. Each daughter cell ends up with a new perfect copy of the DNA, in which one of the strands came from the original DNA molecule and the other was newly synthesised.

Even in nature, in a system which has evolved over billions of years, nothing is perfect and occasionally the replication machinery makes a mistake. It might try to insert a T where a C should really go. When this happens the error is almost always repaired very quickly by another set of proteins that can recognise that this has happened, take out the wrong base and put in the right one. This is the DNA repair machinery, and one of the reasons it's able to act is because when the wrong bases pair up, it recognises that the DNA 'zip' isn't done up properly.

The cell puts a huge amount of energy into keeping the DNA copies completely faithful to the original template. This makes

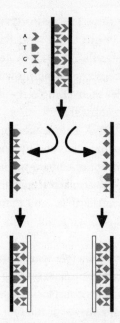

Figure 3.2 The first stage in replication of DNA is the separation of the two strands of the double helix. The bases on each separated backbone act as the template for the creation of a new strand. This ensures that the two new double-stranded DNA molecules have exactly the same base sequence as the parent molecule. Each new double helix of DNA has one backbone that was originally part of the parent molecule (in black) and one freshly synthesised backbone (in white).

sense if we go back to our model of DNA as a script. Consider one of the most famous lines in all of English literature:

> *O Romeo, Romeo! wherefore art thou Romeo?*

If we insert just one extra letter, then no matter how well the line is delivered on stage, its effect is unlikely to be the one intended by the Bard:

> *O Romeo, Romeo! wherefore fart thou Romeo?*

This puerile example illustrates why a script needs to be reproduced faithfully. It can be the same with our DNA – one inappropriate change (a mutation) can have devastating effects. This is particularly true if the mutation is present in an egg or a sperm, as this can ultimately lead to the birth of an individual in whom all the cells carry the mutation. Some mutations have devastating clinical effects. These range from children who age so prematurely that a ten-year-old has the body of a person of 70, to women who are pretty much predestined to develop aggressive and difficult to treat breast cancer before they are 40 years of age. Thankfully, these sorts of genetic mutations and conditions are relatively rare compared with the types of diseases that afflict most people.

The 50,000,000,000,000 or so cells in a human body are all the result of perfect replication of DNA, time after time after time, whenever cells divide after the formation of that single-cell zygote from Chapter 1. This is all the more impressive when we realise just how much DNA has to be reproduced each time one cell divides to form two daughter cells. Each cell contains six billion base-pairs of DNA (half originally came from your father and half from your mother). This sequence of six billion base-pairs is what we call the genome. So every single cell division in the human body was the result of copying 6,000,000,000 bases of DNA. Using the same type of calculation as in Chapter 1, if we count one base-pair every second without stopping, it would take a mere 190 years to count all the bases in the genome of a cell. When we consider that a baby is born just nine months after the creation of the single-celled zygote, we can see that our cells must be able to replicate DNA really fast.

The three billion base-pairs we inherit from each parent aren't formed of one long string of DNA. They are arranged into smaller bundles, which are the chromosomes. We'll delve deeper into these in Chapter 9.

Reading the script

Let's go back to the more fundamental question of what these six billion base-pairs of DNA actually do, and how the script works. More specifically how can a code that only has four letters (A, C, G and T) create the thousands and thousands of different proteins found in our cells? The answer is surprisingly elegant. It could be described as the modular paradigm of molecular biology but it's probably far more useful to think of it as Lego.

Lego used to have a great advertising slogan 'It's a new toy every day', and it was very accurate. A large box of Lego contains a limited number of designs, essentially a fairly small range of bricks of certain shapes, sizes and colours. Yet it's possible to use these bricks to create models of everything from ducks to houses, and from planes to hippos. Proteins are rather like that. The 'bricks' in proteins are quite small molecules called amino acids, and there are twenty standard amino acids (different Lego bricks) in our cells. But these twenty amino acids can be joined together in an incredible array of combinations of all sorts of diversity and length, to create an enormous number of proteins.

That still leaves the problem of how even as few as twenty amino acids can be encoded by just four bases in DNA. The way this works is that the cell machinery 'reads' DNA in blocks of three base-pairs at a time. Each block of three is known as a codon and may be AAA, or GCG or any other combination of A, C, G and T. From just four bases it's possible to create sixty-four different codons, more than enough for the twenty amino acids. Some amino acids are coded for by more than one codon. For example, the amino acid called lysine is coded for by AAA and AAG. A few codons don't code for amino acids at all. Instead they act as signals to tell the cellular machinery that it's at the end of a protein-coding sequence. These are referred to as stop codons.

How exactly does the DNA in our chromosomes act as a script for producing proteins? It does it through an intermediary protein, a molecule called messenger RNA (mRNA). mRNA is very like DNA although it does differ in a few significant details. Its backbone is slightly different from DNA (hence RNA, which stands for ribonucleic acid rather than deoxyribonucleic acid); it is single-stranded (only one backbone); it replaces the T base with a very similar but slightly different one called U (we don't need to go into the reason it does this here). When a particular DNA stretch is 'read' so that a protein can be produced using that bit of script, a huge complex of proteins unzips the right piece of DNA and makes mRNA copies. The complex uses the base-pairing principle to make perfect mRNA copies. The mRNA molecules are then used as temporary templates at specialised structures in the cell that produce protein. These read the three letter codon code and stitch together the right amino acids to form the longer protein chains. There is of course a lot more to it than all this, but that's probably sufficient detail.

An analogy from everyday life may be useful here. The process of moving from DNA to mRNA to protein is a bit like controlling an image from a digital photograph. Let's say we take a photograph on a digital camera of the most amazing thing in the world. We want other people to have access to the image, but we don't want them to be able to change the original in any way. The raw data file from the camera is like the DNA blueprint. We copy it into another format, that can't be changed very much – a PDF maybe – and then we email out thousands of copies of this PDF, to everyone who asks for it. The PDF is the messenger RNA. If people want to, they can print paper copies from this PDF, as many as they want, and these paper copies are the proteins. So everyone in the world can print the image, but there is only one original file.

Why so complicated, why not just have a direct mechanism? There are a number of good reasons that evolution has favoured this indirect method. One of them is to prevent damage to the script, the original image file. When DNA is unzipped it is relatively susceptible to damage and that's something that cells have evolved to avoid. The indirect way in which DNA codes for proteins minimises the period of time for which a particular stretch of DNA is open and vulnerable. The other reason this indirect method has been favoured by evolution is that it allows a lot of control over the amount of a specific protein that's produced, and this creates flexibility.

Consider the protein called alcohol dehydrogenase (ADH). This is produced in the liver and breaks down alcohol. If we drink a lot of alcohol, the cells of our livers will increase the amounts of ADH they produce. If we don't drink for a while, the liver will produce less of this protein. This is one of the reasons why people who drink frequently are better able to tolerate the immediate effects of alcohol than those who rarely drink, who will become tipsy very quickly on just a couple of glasses of wine. The more often we drink alcohol, the more ADH protein our livers produce (up to a limit). The cells of the liver don't do this by increasing the number of copies of the *ADH* gene. They do this by reading the *ADH* gene more efficiently, i.e. producing more mRNA copies and/or by using these mRNA copies more efficiently as protein templates.

As we shall see, epigenetics is one of the mechanisms a cell uses to control the amount of a particular protein that is produced, especially by controlling how many mRNA copies are made from the original template.

The last few paragraphs have all been about how genes encode proteins. How many genes are there in our cells? This seems like a simple question but oddly enough there is no agreed figure on this. This is because scientists can't agree on how to define a gene. It used to be quite straightforward – a gene was a stretch of DNA that encoded a protein. We now know that this is far too simplistic. However, it's certainly true to say that all proteins are encoded by genes, even if not all genes encode proteins. There are about 20,000 to 24,000 protein-encoding genes in our DNA, a much lower estimate than the 100,000 that scientists thought was a good guess just ten years ago[1].

Editing the script

Most genes in human cells have quite a similar structure. There's a region at the beginning called the promoter, which binds the protein complexes that copy the DNA to form mRNA. The protein complexes move along through what's known as the body of the gene, making a long mRNA strand, until they finally fall off at the end of the gene.

Imagine a gene body that is 3,000 base-pairs long, a perfectly sensible length for a gene. The mRNA will also be 3,000 base-pairs long. Each amino acid is encoded by a codon composed of three bases, so we would predict that this mRNA will encode a protein that is 1,000 amino acids long. But, perhaps unexpectedly, what we find is that the protein is usually considerably shorter than this.

If the sequence of a gene is typed out it looks like a long string of combinations of the letters A, C, G and T. But if we analyse this with the right software, we find that we can divide that long string into two types of sequences. The first type is called an exon (for *ex*pressed sequence) and an exon can code for a run of amino acids. The second type is called an intron (for *in*expressed sequence). This doesn't code for a run of amino acids. Instead it contains lots of the 'stop' codons that signal that the protein should come to an end.

When the mRNA is first copied from the DNA it contains the whole run of exons and introns. Once this long RNA molecule has been created, another multi-sub-unit protein complex comes along. It removes all the intron sequences and then joins up the exons to create an mRNA that codes for a continuous run of amino acids. This editing process is called splicing.

This again seems extremely complicated, but there's a very good reason that this complex mechanism has been favoured by evolution. It's because it enables a cell to use a relatively small number of genes to create a much bigger number of proteins. The way this works is shown in Figure 3.3.

The initial mRNA contains all the exons and all the introns. Then it's spliced to remove the introns. But during this splicing some of the exons may also be removed. Some exons will be retained in the final mRNA, others will be skipped over. The various proteins that this creates may have quite similar functions, or

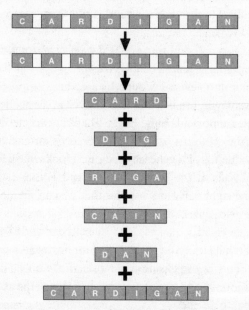

Figure 3.3   The DNA molecule is shown at the very top of this diagram. The exons, which code for stretches of amino acids, are shown in the dark boxes. The introns, which don't code for amino acid sequences, are represented by the white boxes. When the DNA is first copied into RNA, indicated by the first arrow, the RNA contains both the exons and the introns. The cellular machinery then removes some or all of the introns (the process known as splicing). The final messenger RNA molecules can thereby code for a variety of proteins from the same gene, as represented by the various words shown in the diagram. For simplicity, all the introns and exons have been drawn as the same size, but in reality they can vary widely.

they may differ dramatically. The cell can express different proteins depending on what that cell has to do at a particular time, or because of different signals that it receives. If we define a gene as something that encodes a protein, this mechanism means that just 20,000 or so genes can code for far more than just 20,000 proteins.

Whenever we describe the genome we talk about it in very two-dimensional terms, almost like a railway track. Peter Fraser's laboratory at the Babraham Institute outside Cambridge has published some extraordinary work showing it's probably nothing like this at all. He works on the genes that code for the proteins required to make haemoglobin, the pigment in red blood cells that carries oxygen all around the body. There are a number of different proteins needed to create the final pigment, and they lie on different chromosomes. Doctor Fraser has shown that in cells that produce large amounts of haemoglobin, these chromosome regions become floppy and loop out like tentacles sticking out of the body of an octopus. These floppy regions mingle together in a small area of the cell nucleus, waving about until they can find each other. By doing this, there is an increased chance that all the proteins needed to create the functional haemoglobin pigment will be expressed together at the same time[2].

Each cell in our body contains 6,000,000,000 base-pairs. About 120,000,000 of these code for proteins. One hundred and twenty million sounds like a lot, but it's actually only 2 per cent of the total amount. So although we think of proteins as being the most important things our cells produce, about 98 per cent of our genome doesn't code for protein.

Until recently, the reason that we have so much DNA when so little of it leads to a protein was a complete mystery. In the last ten years we've finally started to get a grip on this, and once again it's connected with regulating gene expression through epigenetic mechanisms. It's now time to move on to the molecular biology of epigenetics.

Chapter 4

Life As We Know It Now

*The important thing in science is not so much to obtain
new facts as to discover new ways of thinking about them.*
Sir William Bragg

So far this book has focused mainly on outcomes, the things that
we can observe that tell us that epigenetic events happen. But
every biological phenomenon has a physical basis and that's what
this chapter is about. The epigenetic outcomes we've described
are all a result of variations in expression of genes. The cells of
the retina express a different set of genes from the cells in the
bladder, for example. But how do the different cell types switch
different sets of genes on or off?

The specialised cell types in the retina and in the bladder are
each at the bottom of one of the troughs in Waddington's epigenetic
landscape. The work of both John Gurdon and Shinya Yamanaka
showed us that whatever mechanism cells use for staying in these
troughs, it's not anything to do with changing the DNA blueprint
of the cell. That remains intact and unchanged. Therefore keeping
specific sets of genes turned on or off must happen through some
other mechanism, one that can be maintained for a really long time.
We know this must be the case because some cells, like the neurons
in our brains, are remarkably long-lived. The neurons in the brain
of an 85-year-old person, for example, are about 85 years of age.
They formed when the individual was very young, and then stayed
the same for the rest of their life.

But other cells are different. The top layer of skin cells, the epi-
dermis, is replaced about every five weeks, from constantly dividing stem cells in the deeper layers of that tissue. These stem cells

always produce new skin cells, and not, for example, muscle cells. Therefore the system that keeps certain sets of genes switched on or off must also be a mechanism that can be passed on from parent cell to daughter cell every time there is a cell division.

This creates a paradox. Researchers have known since the work of Oswald Avery and colleagues in the mid-1940s that DNA is the material in cells that carries our genetic information. If the DNA stays the same in different cell types in one individual, how can the incredibly precise patterns of gene expression be transmitted down through the generations of cell division?

Our analogy of actors reading a script is again useful. Baz Luhrmann hands Leonardo DiCaprio Shakespeare's script for *Romeo and Juliet*, on which the director has written or typed various notes – directions, camera placements and lots of additional technical information. Whenever Leo's copy of the script is photocopied, Baz Luhrmann's additional information is copied along with it. Claire Danes also has the script for *Romeo and Juliet*. The notes on her copy are different from those on her co-star's, but will also survive photocopying. That's how epigenetic regulation of gene expression occurs – different cells have the same DNA blueprint (the original author's script) but carrying varied molecular modifications (the shooting script) which can be transmitted from mother cell to daughter cell during cell division.

These modifications to DNA don't change the essential nature of the A, C, G and T alphabet of our genetic script, our blueprint. When a gene is switched on and copied to make mRNA, that mRNA has exactly the same sequence, controlled by the base-pairing rules, irrespective of whether or not the gene is carrying an epigenetic addition. Similarly, when the DNA is copied to form new chromosomes for cell division, the same A, C, G and T sequences are copied.

Since epigenetic modifications don't change what a gene codes for, what do they do? Basically, they can dramatically change how well a gene is expressed, or if it is expressed at all. Epigenetic modifications can also be passed on when a cell divides, so this provides a mechanism for how control of gene expression stays consistent from mother cell to daughter cell. That's why skin stem cells only give rise to more skin cells, not to any other cell type.

Sticking a grape on DNA

The first epigenetic modification to be identified was DNA methylation. Methylation means the addition of a methyl group to another chemical, in this case DNA. A methyl group is very small. It's just one carbon atom linked to three hydrogen atoms. Chemists describe atoms and molecules by their 'molecular weight', where the atom of each element has a different weight. The average molecular weight of a base-pair is around 600 Da (the Da stands for Daltons, the unit that is used for molecular weight). A methyl group only weighs 15 Da. By adding a methyl group the weight of the base-pair is only increased by 2.5 per cent. A bit like sticking a grape on a tennis ball.

Figure 4.1 shows what DNA methylation looks like chemically.

The base shown is C – cytosine. It's the only one of the four DNA bases that gets methylated, to form 5-methylcytosine. The '5' refers to the position on the ring where the methyl is added, not to the number of methyl groups; there's always only one of these. This methylation reaction is carried out in our cells, and

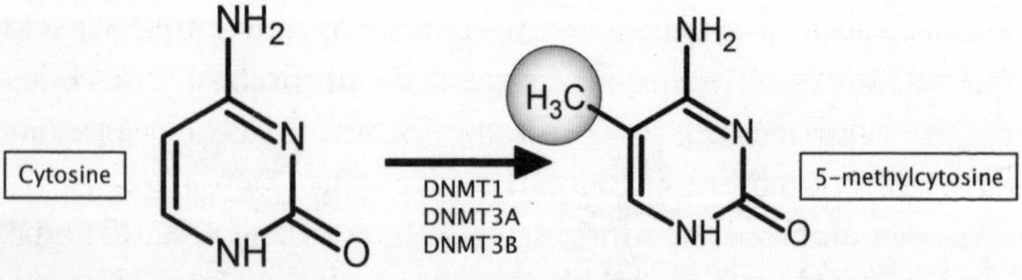

Figure 4.1   The chemical structures of the DNA base cytosine and its epigenetically modified form, 5-methylcytosine. C: carbon; H: hydrogen; N: nitrogen; O: oxygen. For simplicity, some carbon atoms have not been explicitly shown, but are present where there is a junction of two lines.

those of most other organisms, by one of three enzymes called DNMT1, DNMT3A or DNMT3B. DNMT stands for <u>DNA</u> <u>m</u>ethyl<u>t</u>ransferase. The DNMTs are examples of epigenetic 'writers' – enzymes that create the epigenetic code. Most of the time these enzymes will only add a methyl group to a C that is followed by a G. C followed by G is known as CpG.

This CpG methylation is an epigenetic modification, which is also known as an epigenetic mark. The chemical group is 'stuck onto' DNA but doesn't actually alter the underlying genetic sequence. The C has been decorated rather than changed. Given that the modification is so small, it's perhaps surprising that it will come up over and over again in this book, and in any discussion of epigenetics. This is because methylation of DNA has profound effects on how genes are expressed, and ultimately on cellular, tissue and whole-body functions.

In the early 1980s it was shown that if you injected DNA into mammalian cells, the amount of methylation on the injected DNA affected how well it was transcribed into RNA. The more methylated the injected DNA was, the less transcription that occurred[1]. In other words, high levels of DNA methylation were associated with genes that were switched off. However, it wasn't clear how significant this was for the genes normally found in the nuclei of cells, rather than ones that were injected into cells.

The key work in establishing the importance of methylation in mammalian cells came out of the laboratory of Adrian Bird, who has spent most of his scientific career in Edinburgh, Conrad Waddington's old stomping ground. Professor Bird is a Fellow of the Royal Society and a former Governor of the Wellcome Trust, the enormously influential independent funding agency in UK science. He is one of those traditional British scientific types – understated, soft-spoken, non-flashy and drily funny. His lack of self-promotion is in contrast to his stellar international reputation, where he is widely acknowledged as the godfather of DNA methylation and its role in controlling gene expression.

In 1985 Adrian Bird published a key paper in *Cell* showing that most CpG motifs were not randomly distributed throughout the genome. Instead the majority of CpG pairs were concentrated just upstream of certain genes, in the promoter region[2]. Promoters are the stretches of the genome where the DNA transcription complexes bind and start copying DNA to form RNA. Regions where there is a high concentration of CpG motifs are called CpG islands.

In about 60 per cent of the genes that code for proteins, the promoters lie within CpG islands. When these genes are active, the levels of methylation in the CpG island are low. The CpG islands tend to be highly methylated only when the genes are switched off. Different cell types express different genes, so unsurprisingly the patterns of CpG island methylation are also different across different cell types.

For quite some time there was considerable debate about what this association meant. It was the old cause or effect debate. One interpretation was that DNA methylation was essentially a historical modification – genes were repressed by some unknown mechanism and then the DNA became methylated. In this model, DNA methylation was just a downstream consequence of gene repression. The other interpretation was that the CpG island became methylated, and it was this methylation that switched the gene off. In this model the epigenetic modification actually causes the change in gene expression. Although there is still the occasional argument about this among competing labs, the vast majority of scientists in this field now believe that the data generated in the quarter of a century since Adrian Bird's paper are consistent with the second, causal model. Under most circumstances, methylation of the CpG island at the start of a gene turns that gene off.

Adrian Bird went on to investigate how DNA methylation switches genes off. He showed that when DNA is methylated, it binds a protein called MeCP2 (Methyl CpG binding protein 2)[3].

However, this protein won't bind to unmethylated CpG motifs, which is pretty amazing when we look back at Figure 4.1 and think how similar the methylated and unmethylated forms of cytosine really are. The enzymes that add the methyl group to DNA have been described as writers of the epigenetic code. MeCP2 doesn't add any modifications to DNA. Its role is to enable the cell to interpret the modifications on a DNA region. MeCP2 is an example of a 'reader' of the epigenetic code.

Once MeCP2 binds to 5-methylcytosine in a gene promoter it seems to do a number of things. It attracts other proteins that also help to switch the gene off[4]. It may also stop the DNA transcription machinery from binding to the gene promoter, and this prevents mRNA messenger molecule from being produced[5]. Where genes and their promoters are very heavily methylated, binding of MeCP2 seems to be part of a process where that region of a chromosome gets shut down almost permanently. The DNA becomes incredibly tightly coiled up and the gene transcription machinery can't get access to the base-pairs to make mRNA copies.

This is one of the reasons why DNA methylation is so important. Remember those 85 year old neurons in the brains of senior citizens? For over eight decades DNA methylation has kept certain regions of the genome incredibly tightly compacted and so the neuron has kept certain genes completely repressed. This is why our brain cells never produce haemoglobin, for example, or digestive enzymes.

But what about the other situation, the example of skin stem cells dividing very frequently but always just creating new skin cells, rather than some other cell type such as bone? In this situation, the pattern of DNA methylation is passed from mother cell to daughter cells. When the two strands of the DNA double helix separate, each gets copied using the base-pairing principle, as we saw in Chapter 3. Figure 4.2 illustrates what happens when this replication occurs in a region where the CpG is methylated on the C.

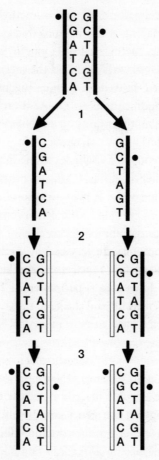

Figure 4.2   This schematic shows how DNA methylation patterns can be preserved when DNA is replicated. The methyl group is represented by the black circle. Following separation of the parent DNA double helix in step 1, and replication of the DNA strands in step 2, the new strands are 'checked' by the DNA methyltransferase 1 (DNMT1) enzyme. DNMT1 can recognise that a methyl group at a cytosine motif on one strand of a DNA molecule is not matched on the newly synthesised strand. DNMT1 transfers a methyl group to the cytosine on the new strand (step 3). This only occurs where a C and a G are next to each other in a CpG motif. This process ensures that the DNA methylation patterns are maintained following DNA replication and cell division.

DNMT1 can recognise if a CpG motif is only methylated on one strand. When DNMT1 detects this imbalance, it replaces the 'missing' methylation on the newly copied strand. The daughter cells will therefore end up with the same DNA methylation patterns as the parent cell. As a consequence, they will repress the same genes as the parent cell and the skin cells will stay as skin cells.

Miracle mice on *YouTube*

Epigenetics has a tendency to crop up in places where scientists really aren't expecting it. One of the most interesting examples of this in recent years has related to MeCP2, the protein that reads the DNA methylation mark. Several years ago, the now discredited theory of the MMR vaccine causing autism was at its height, and getting lots of coverage in the general media. One very respected UK broadsheet newspaper covered in depth the terribly sad story of a little girl. As a baby she initially met all the usual developmental milestones. Shortly after receiving an MMR jab not long before her first birthday she began to deteriorate rapidly, losing most of the skills she had gained. By the time the journalist wrote the article, the little girl was about four years old and was described as having the most severely autistic symptoms the author had ever seen. She had not developed language, appeared to have very severe learning difficulties and her actions were very limited and repetitive, with very few purposeful hand actions (she no longer reached out for food, for example). Development of this incredibly severe disability was undoubtedly a tragedy for her and for her family.

But if a reader with any sort of background in neurogenetics read this article, two things probably struck them immediately. The first was that it's very unusual – not unheard of but pretty uncommon – for girls to present with such severe autism. This is much more common in boys. The second thing that would have struck them was that this case sounded exactly the same as a rare genetic disorder called Rett syndrome, right down to the normal early development and the timing and types of symptoms. It's just coincidence that the symptoms of Rett syndrome, and indeed of most types of autism, first start becoming obvious at around the same age as when infants are typically given the MMR vaccination.

But what does this have to do with epigenetics? In 1999, a group led by the eminent neurogeneticist Huda Zoghbi at the Baylor College of Medicine in Houston, Texas showed that the majority of cases of Rett syndrome are caused by mutations in *MeCP2*, the gene which encodes the reader of methylated DNA. The children with this disorder have a mutation in the *MeCP2* gene which means that they don't produce a functional MeCP2 protein. Although their cells are perfectly capable of methylating DNA correctly, the cells can't read this part of the epigenetic code properly.

The severe clinical symptoms of children with the *MeCP2* mutation tell us that reading the epigenetic code properly is very important. But they also tell us other things. Not all the tissues of girls with Rett syndrome are equally affected, so perhaps this particular epigenetic pathway is more important in some tissues than others. Because the girls develop severe mental retardation, we can deduce that having the right amount of normal MeCP2 protein is really important in the brain. Given that these children seem to be fairly unaffected in other tissues such as liver or kidney, perhaps MeCP2 activity isn't as important in these tissues. It could be that DNA methylation itself isn't so critical in these organs, or maybe these tissues contain other proteins in addition to MeCP2 that can read this part of the epigenetic code.

Long-term, scientists, physicians and families of children with Rett syndrome would dearly love to be able to use our increased understanding of the disease to help us find better treatments. This is a huge challenge, as we would be trying to intervene in a condition that affects the brain as a result of a gene mutation that is present throughout development, and beyond.

One of the most debilitating aspects of Rett syndrome is the profound mental retardation that is an almost universal symptom. Nobody knew if it would be possible to reverse a neurodevelopmental problem such as mental retardation once it had become established, but the general feeling about this wasn't optimistic. Adrian Bird remains a major figure in our story. In 2007 he published an astonishing paper in *Science*, in which he and his colleagues showed that Rett syndrome could be reversed, in a mouse model of the disease.

Adrian Bird and his colleagues created a cloned strain of mice in which the *Mecp2* gene was inactivated. They used the types of technologies pioneered by Rudolf Jaenisch. These mice developed severe neurological symptoms, and as adults they exhibited hardly any normal mouse activities. If you put a normal mouse in the middle of a big white box, it will almost immediately begin to explore its surroundings. It will move around a lot, it will tend to follow the edges of the box just like a normal house mouse scurrying along by the skirting boards, and it will frequently rear up on its back legs to get a better view. A mouse with the *Mecp2* mutation does very few of these things – put it in the middle of a big white box and it will tend to stay there.

When Adrian Bird created his mouse strain with the *Mecp2* mutation, he also engineered it so that the mice would also be carrying a normal copy of *Mecp2*. However, this normal copy was silent – it wasn't switched on in the mouse cells. The really clever bit of this experiment was that if the mice were given a specific harmless chemical, the normal *Mecp2* gene became activated. This allowed the experimenters to let the mice develop and grow up with no Mecp2 in their cells, and then at a time of the scientists' choosing, the *Mecp2* gene could be switched on.

The results of switching on the *Mecp2* gene were extraordinary. Mice which previously just sat in the middle of the white box suddenly turned into the curious explorers that mice should be[6]. You can find clips of this on *YouTube*, along with interviews with Adrian Bird where he basically concedes that he really never expected to see anything so dramatic[7].

The reason this experiment is so important is that it offers hope that we may be able to find new treatments for really complex neurological conditions. Prior to the publication of this *Science* paper, there had been an assumption that once a complex neurological condition has developed, it is impossible to reverse it. This was especially presumed to be the case for any condition that arises developmentally, i.e. in the womb or in early infancy. This is a critical period when the mammalian brain is making so many of the connections and structures that are used throughout the rest of life. The results from the *Mecp2* mutant mice suggest that in Rett syndrome, maybe all the bits of cellular machinery that are required for normal neurological function are still there in the brain – they just need to be activated properly. If this holds true for humans (and at a brain level we aren't really *that* different from mice) this offers hope that maybe we can start to develop therapies to reverse conditions as complex as mental retardation. We can't do this the way it was done in the mouse, as that was a genetic approach that can only be used in experimental animals and not in humans, but it suggests that it is worth trying to develop suitable drugs that have a similar effect.

DNA methylation is clearly really important. Defects in reading DNA methylation can lead to a complex and devastating neurological disorder that leaves children with Rett syndrome severely disabled throughout their lives. DNA methylation is also essential for maintaining the correct patterns of gene expression in different cell types, either for several decades in the case of our long-lived neurons, or in all daughters of a stem cell in a constantly-replaced tissue such as skin.

But we still have a conceptual problem. Neurons are very different from skin cells. If both cells types use DNA methylation to switch off certain genes, and to keep them switched off, they must be using the methylation at different sets of genes. Otherwise they would all be expressing the same genes, to the same extent, and they would inevitably then be the same types of cells instead of being neurons and skin cells.

The solution to how two cell types can use the same mechanism to create such different outcomes lies in how DNA methylation gets targeted to different regions of the genome in different cell types. This takes us into the second great area of molecular epigenetics. Proteins.

DNA has a friend

DNA is often described as if it's a naked molecule, i.e. DNA and nothing else. If we visualise it at all in our minds, a DNA double helix probably looks like a very long twisty railway track. This is pretty much how we described it in the previous chapter. But in reality it's actually nothing like that, and many of the great breakthroughs in epigenetics came about when scientists began to appreciate this fully.

DNA is intimately associated with proteins, and in particular with proteins called histones. At the moment most attention in epigenetics and gene regulation is focused on four particular histone proteins called H2A, H2B, H3 and H4. These histones have a structure known as 'globular', as they are folded into compact ball-like shapes. However, each also has a loose floppy chain of amino acids that sticks out of the ball, which is called the histone tail. Two copies of each of these four histone proteins come together to form a tight structure called the histone octamer (so called because it's formed of eight individual histones).

It might be easiest to think of this octamer as eight ping-pong balls stacked on top of each other in two layers. DNA coils tightly around this protein stack like a long liquorice whip around marshmallows, to form a structure called the nucleosome. One hundred and forty seven base-pairs of DNA coil around each nucleosome. Figure 4.3 is a very simplified representation of the structure of a nucleosome, where the white strand is DNA and the grey wiggles are the histone tails.

If we had read anything about histones even just fifteen years ago, they would probably have been described as 'packaging proteins', and left at that. It's certainly true that DNA has to be packaged. The nucleus of a cell is usually only about 10 microns in diameter – that's 1/100th of a millimetre – and if the DNA in a cell was just left all floppy and loose it could stretch for 2 metres. The DNA is curled tightly around the histone octamers and these are all stacked closely on top of each other.

Certain regions of our chromosomes have an extreme form of that sort of structure almost all the time. These tend to be regions that don't really code for any genes. Instead, they are structural regions such as the very ends of chromosomes, or areas that are important for separating chromosomes after DNA has been duplicated for cell division.

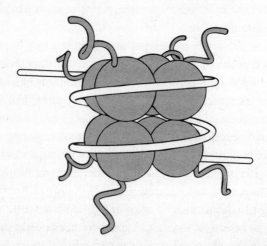

Figure 4.3   The histone octamer (2 molecules each of histones H2A, H2B, H3 and H4) stacked tightly together, and with DNA wrapped around it, forms the basic unit of chromatin called the nucleosome.

The regions of DNA that are really heavily methylated also have this hyper-condensed structure and the methylation is very important in establishing this configuration. It's one of the mechanisms used to keep certain genes switched off for decades in long-lived cell types such as neurons.

But what about those regions that aren't screwed down tight, where there are genes that are switched on or have the potential to be switched on? This is where the histones really come into play. There is so much more to histones than just acting as a molecular reel for wrapping DNA around. If DNA methylation represents the semi-permanent additional notes on our script of *Romeo and Juliet*, histone modifications are the more tentative additions. They may be like pencil marks, that survive a few rounds of photocopying but eventually fade out. They may be even more transient, like Post-It notes, used very temporarily.

A substantial number of the breakthroughs in this field have come from the lab of Professor David Allis at Rockefeller University in New York. He's a trim, neat, clean-shaven American who looks much younger than his 60 years and is exceptionally popular amongst his peers. Like many epigeneticists, he began his career in the field of developmental biology. Just like Adrian Bird, and John Gurdon before him, David Allis wears his stellar reputation in epigenetics very lightly. In a remarkable flurry of papers in 1996, he and his colleagues showed that histone proteins were chemically modified in cells, and that this modification increased expression of genes near a specific modified nucleosome[8].

The histone modification that David Allis identified was called acetylation. This is the addition of a chemical group called an acetyl, in this case to a specific amino acid named lysine on the floppy tail of one of the histones. Figure 4.4 shows the structures of lysine and acetyl-lysine, and we can again see that the modification is relatively small. Like DNA methylation, lysine acetylation is an epigenetic mechanism for altering gene expression which doesn't change the underlying gene sequence.

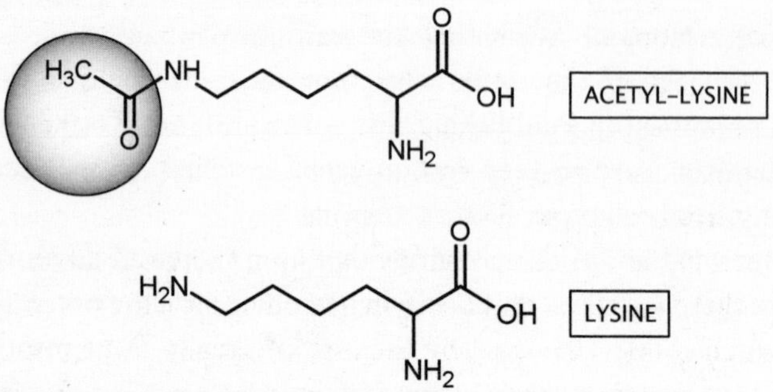

Figure 4.4 The chemical structures of the amino acid lysine and its epigenetically modified form, acetyl-lysine. C: carbon; H: hydrogen; N: nitrogen; O: oxygen. For simplicity, some carbon atoms have not been explicitly shown, but are present where there is a junction of two lines.

So back in 1996 there was a nice simple story. DNA methylation turned genes off and histone acetylation turned genes on. But gene expression is much more subtle than genes being either on or off. Gene expression is rarely an on-off toggle switch; it's much more like the volume dial on a traditional radio. So perhaps it was unsurprising that there turned out to be more than one histone modification. In fact, more than 50 different epigenetic modifications to histone proteins have been identified since David Allis's initial work, both by him and by a large number of other laboratories[9]. These modifications all alter gene expression but not always in the same way. Some histone modifications push gene expression up, others drive it down. The pattern of modifications is referred to as a histone code[10]. The problem that epigeneticists face is that this is a code that is extraordinarily difficult to read.

Imagine a chromosome as the trunk of a very big Christmas tree. The branches sticking out all over the tree are the histone tails and these can be decorated with epigenetic modifications. We pick up the purple baubles and we put one, two or three purple baubles on some of the branches. We also have green icicle decorations and we can put either one or two of these on some branches, some of which already have purple baubles on them. Then we pick up the red stars but are told we can't put these on a branch if the adjacent branch has any purple baubles. The gold snowflakes and green icicles can't be present on the same branch. And so it goes on, with increasingly complex rules and patterns. Eventually, we've used all our decorations and we wind the lights around the tree. The bulbs represent individual genes. By a magical piece of software programming, the brightness of each bulb is determined by the precise conformation of the decorations surrounding it. The likelihood is that we would really struggle to predict the brightness of most of the bulbs because the pattern of Christmas decorations is so complicated.

That's where scientists currently are in terms of predicting how all the various histone modification combinations work together to influence gene expression. It's reasonably clear in many cases what individual modifications can do, but it's not yet possible to make accurate predictions from complex combinations.

There are major efforts being made to learn how to understand this code, with multiple labs throughout the world collaborating or competing in the use of the fastest and most complex technologies to address this problem. The reason for this is that although we may not be able to read the code properly yet, we know enough about it to understand that it's extremely important.

Build a better mousetrap

Some of the key evidence comes from developmental biology, the field from which so many great epigenetic investigators have emerged. As we have already described, the single-celled zygote divides, and very quickly daughter cells start to take on discrete functions. The first noticeable event is that the cells of the early embryo split into the inner cell mass (ICM) and the trophoectoderm. The ICM cells in particular start to differentiate to form an increasing number of different cell types. This rolling of the cells down the epigenetic landscape is, to quite a large degree, a self-perpetuating system.

The key concept to grasp at this stage is the way that waves of gene expression and epigenetic modifications follow on from each other. A useful analogy for this is the game of *Mousetrap*, first produced in the early 1960s and still on sale today. Players have to build an insanely complex mouse trap during the course of the game. The trap is activated at one end by the simple act of releasing a ball. This ball passes down and through all sorts of contraptions including a slide, a kicking boot, a flight of steps and a man jumping off a diving board. As long as the pieces have been put together properly, the whole ridiculous cascade operates perfectly, and the toy mice get caught under a net. If one of the pieces is just slightly mis-aligned, the crazy sequence judders to a halt and the trap doesn't work.

The developing embryo is like *Mousetrap*. The zygote is pre-loaded with certain proteins, mainly from the egg cytoplasm. These egg-derived proteins move into the nucleus and bind to target genes, which we'll call *Boots* (in honour of *Mousetrap*), and regulate their expression. They also attract a select few epigenetic enzymes to the *Boots* genes. These epigenetic enzymes may also have been 'donated' from the egg cytoplasm and they set up longer-lasting modifications to the DNA and histone proteins of chromatin, also influencing how these *Boots* genes are switched on or off. The *Boots* proteins bind to the *Divers* genes, and switch these on. Some of these *Divers* genes may themselves encode epigenetic enzymes, which will form complexes on members of the *Slides* family of genes, and so on. The genetic and epigenetic proteins work together in a seamless orderly procession, just like the events in *Mousetrap* once the ball has been released. Sometimes a cell will express a little more or a little less of a key factor, one whose expression is on a finely balanced threshold. This has the potential to alter the developmental path that the cell takes, as if twenty *Mousetrap* games had been connected up. Slight deviations in how the pieces were fitted together, or how the ball rolled at critical moments, would trigger one trap and not another.

The names in our analogy are made up, but we can apply this to a real example. One of the key proteins in the very earliest stages of embryonic development is Oct4. Oct4 protein binds to certain key genes, and also attracts a specific epigenetic enzyme. This enzyme modifies the chromatin and alters the regulation of that gene. Both Oct4 and the epigenetic enzyme with which it works are essential for development of the early embryo. If either is absent, the zygote can't even develop as far as creating an ICM.

The patterns of gene expression in the early embryo eventually feed back on themselves. When certain proteins are expressed, they can bind to the *Oct4* promoter and switch off expression of this gene. Under normal circumstances, somatic cells just don't express Oct4. It would be too dangerous for them to do so because Oct4 could disrupt the normal patterns of gene expression in differentiated cells, and make them more like stem cells.

This is exactly what Shinya Yamanaka did when he used Oct4 as a reprogramming factor. By artificially creating very high levels of Oct4 in differentiated cells, he was able to 'fool' the cells into acting like early developmental cells. Even the epigenetic modifications were reset – that's how powerful this gene is.

Normal development has yielded important evidence of the significance of epigenetic modifications in controlling cell fate. Cases where development goes awry have also shown us how important epigenetics can be.

For example, a 2010 publication in *Nature Genetics* identified the mutations that cause a rare disease called Kabuki syndrome. Kabuki syndrome is a complex developmental disorder with a range of symptoms that include mental retardation, short stature, facial abnormalities and cleft palate. The paper showed that Kabuki syndrome is caused by mutations in a gene called *MLL2*[11]. The MLL2 protein is an epigenetic writer that adds methyl groups to a specific lysine amino acid at position 4 on

. Patients with this mutation are unable to write their
code properly, and this leads to their symptoms.

Human diseases can also be caused by mutations in enzymes
that remove epigenetic modifications, i.e. 'erasers' of the epigenetic code. Mutations in a gene called *PHF8*, which removes
methyl groups from a lysine at position 20 on histone H3, cause a
syndrome of mental retardation and cleft palate[12]. In these cases,
the patient's cells put epigenetic modifications on without problems, but don't remove them properly.

It's interesting that although the MLL2 and PHF8 proteins
have different roles, the clinical symptoms caused by mutations
in these genes have overlaps in their presentation. Both lead to
cleft palate and mental retardation. Both of these symptoms are
classically considered as reflecting problems during development.
Epigenetic pathways are important throughout life, but seem to
be particularly significant during development.

In addition to these histone writers and erasers there are over
100 proteins that act as 'readers' of this histone code by binding to epigenetic marks. These readers attract other proteins and
build up complexes that switch on or turn off gene expression.
This is similar to the way that MeCP2 helps turn off expression
of genes that are carrying DNA methylation.

Histone modifications are different to DNA methylation in a
very important way. DNA methylation is a very stable epigenetic
change. Once a DNA region has become methylated it will tend to
stay methylated under most conditions. That's why this epigenetic
modification is so important for keeping neurons as neurons, and
why there are no teeth in our eyeballs. Although DNA methylation *can* be removed in cells, this is usually only under very specific circumstances and it's quite unusual for this to happen.

Most histone modifications are much more plastic than this.
A specific modification can be put on a histone at a particular
gene, removed and then later put back on again. This happens in
response to all sorts of stimuli from outside the cell nucleus. The

stimuli can vary enormously. In some cell types the histone code may change in response to hormones. These include insulin signalling to our muscle cells, or oestrogen affecting the cells of the breast during the menstrual cycle. In the brain the histone code can change in response to addictive drugs such as cocaine, whereas in the cells lining the gut, the pattern of epigenetic modifications will alter depending on the amounts of fatty acids produced by the bacteria in our intestines. These changes in the histone code are one of the key ways in which nurture (the environment) interacts with nature (our genes) to create the complexity of every higher organism on earth.

Histone modifications also allow cells to 'try out' particular patterns of gene expression, especially during development. Genes become temporarily inactivated when repressive histone modifications (those which drive gene expression down) are established on the histones near those genes. If there is an advantage to the cell in those genes being switched off, the histone modifications may last long enough to lead to DNA methylation. The histone modifications attract reader proteins that build up complexes of other proteins on the nucleosome. In some cases the complexes may include DNMT3A or DNMT3B, two of the enzymes that deposit methyl groups on CpG DNA motifs. Under these circumstances, the DNMT3A or 3B can 'reach across' from the complex on the histone and methylate the adjacent DNA. If enough DNA methylation takes place, expression of the gene will shut down. In extreme circumstances the whole chromosome region may become hyper-compacted and inactivated for multiple cell divisions, or for decades in a non-dividing cell like a neuron.

Why have organisms evolved such complex patterns of histone modifications to regulate gene expression? The systems seem particularly complex when you contrast them with the fairly all-or-nothing effects of DNA methylation. One of the reasons is probably because the complexity allows sophisticated fine-tuning of gene expression. Because of this, cells and organisms can adapt their gene expression appropriately in response to changes in their environment, such as availability of nutrients or exposure to viruses. But as we shall see in the next chapter, this fine-tuning can result in some very strange consequences indeed.

Chapter 5

Why Aren't Identical Twins Actually Identical?

There are two things in life for which we are never prepared: twins.
Josh Billings

Identical twins have been a source of fascination in human cultures for millennia, and this fascination continues right into the present day. Just taking Western European literature as one source, we can find the identical twins Menaechmus and Sosicles in a work of Plautus from around 200 B.C.; the re-working of the same story by Shakespeare in *The Comedy of Errors*, written around 1590; Tweedledum and Tweedledee in Lewis Carroll's *Through the Looking-Glass, and What Alice Found There* written in 1871; right up to the Weasley twins in the *Harry Potter* novels of J. K. Rowling. There is something inherently intriguing about two people who seem exactly the same as one another.

But there is something that interests all of us even more than the extraordinary similarities of identical twins, and that is when we can see their differences. It's a device that's been repeatedly used in the arts, from Frederic and Hugo in Jean Anhouil's *Ring around the Moon* to Beverley and Elliott Mantle in David Cronenberg's *Dead Ringers*. Taking this to its extreme you could even cite Dr Jekyll and his alter ego Mr Hyde, the ultimate 'evil twin'. The differences between identical twins have certainly captured the imaginations of creative people from all branches of the arts, but they have also completely captivated the world of science.

The scientific term for identical twins is monozygotic (MZ) twins. They were both derived from the same single-cell zygote formed from the fusion of one egg and one sperm. In the case of MZ twins the inner cell mass of the blastocyst split into two during the early cell divisions, like slicing a doughnut in half, and gave rise to two embryos. And these embryos are *genetically* identical.

This splitting of the inner cell mass to form two separate embryos is generally considered a random event. This is consistent with the frequency of MZ twins being pretty much the same throughout all human populations, and with the fact that identical twins don't run in families. We tend to think of MZ twins as being very rare but this isn't really the case. About one in every 250 full-term pregnancies results in the birth of a pair of MZ twins, and there are around ten million pairs of identical twins around the world today.

MZ twins are particularly fascinating because they help us to determine the degree to which genetics is the driving force for life events such as particular illnesses. They basically allow us to explore mathematically the link between the sequences of our genes (genotype) and what we are like (phenotype), be this in terms of height, health, freckles or anything else we would like to measure. This is done by calculating how often both twins in a pair present with the same disease. The technical term for this is the concordance rate.

Achondroplasia, a relatively common form of short-limbed dwarfism, is an example of a condition in which MZ twins are almost invariably affected in the same way. If one twin has achondroplasia, so does the other one. The disease is said to show 100 per cent concordance. This isn't surprising as achondroplasia is caused by a specific genetic mutation. Assuming that the mutation was present in either the egg or the sperm that fused to form the zygote, all the daughter cells that form the inner cell mass and ultimately the two embryos will also carry the mutation.

However, relatively few conditions show 100 per cent concordance, as the majority of illnesses are not caused by one overwhelming mutation in a key gene. This creates the problem of how to determine if genetics plays a role, and if so, how great this role is. This is where twin studies have become so valuable. If we study large groups of MZ twins we can determine what percentage of them is concordant or discordant for a particular condition. If one twin has a disease, does the other twin also tend to develop it as well?

Figure 5.1 is a graph showing concordance rates for schizophrenia. This shows that the more closely related we are to someone with this disease, the more likely we are to develop it ourselves. The most important parts of the graph to look at are the two bars at the bottom, which deal with twins. From this we can compare the concordance rates for identical and non-identical (fraternal) twins. Non-identical twins share the same developmental environment (the uterus) but genetically are no more similar than any

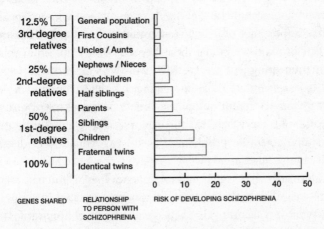

Figure 5.1 The concordance rates for schizophrenia. The more genetically related two individuals are, the more likely it is that if one individual has the disease, their relative will also develop the disorder. However, even in genetically identical monozygotic twins, the concordance rate for schizophrenia does not reach 100 per cent. Data taken from *The Surgeon General's Report on Mental Health* http://www.surgeongeneral.gov/library/mentalhealth/chapter4/sec4_1.html#etiology

other pair of siblings, as they arose from two separate zygotes as a consequence of the fertilisation of two eggs. The comparison between the two types of twins is important because generally speaking, the twins in a pair (whether identical or non-identical) are likely to have shared pretty similar environments. If schizophrenia was caused mainly by environmental factors, we would expect the concordance rates for the disease to be fairly similar between identical and non-identical twins. Instead, what we see is that in non-identical twins, if one twin develops schizophrenia, the other twin has a 17 per cent chance of doing the same. But in MZ twins this risk jumps to nearly 50 per cent. The almost three-fold higher risk for identical versus non-identical twins tells us that there is a major genetic component to schizophrenia.

Similar studies have shown that there is also a substantial genetic component to a significant number of other human disorders, including multiple sclerosis, bipolar disorder, systemic lupus erythematosus and asthma. This has been really useful in understanding the importance of genetic susceptibility to complex diseases.

But in many ways, it's the other side of the question that is more interesting. It's not the MZ twins who both develop a specific disease who are most interesting. It's the MZ twins who end up with very different outcomes – one a paranoid schizophrenic, one mentally very healthy, for example – who create the most intriguing scientific problem. Why do two genetically identical individuals, who in many cases have experienced very similar environments, have such variable phenotypes? Similarly, why is it quite rare for both MZ twins in a pair to develop type 1 diabetes? What is it, in addition to the genetic code, that governs these health outcomes?

How epigenetics drives a wedge between twins

One possible explanation would be that quite randomly the twin with schizophrenia had spontaneously developed mutations in genes in certain cells, for example in the brain. This could happen if the DNA replication machinery had malfunctioned at some point during brain development. These changes might increase his or her susceptibility to a disorder. This is theoretically possible, but scientists have failed to find much data to support this theory.

Of course, the standard answer has always been that discordancy between the twins is due to differences in their environments. Sometimes this is clearly true. If we were monitoring longevity, for example, one twin getting knocked over and killed by a number 47 bus would certainly represent an environmental difference. But this is an extreme scenario. Many twins share a fairly similar environment, especially in early development. Even so, it is certainly possible that there are multiple subtle environmental differences that may be hard to monitor appropriately.

But if we invoke the environment as the other important factor in development of disease, this raises another problem. It still leaves the question of how the environment does this. Somehow the environmental stimuli – be these compounds in our food, chemicals in cigarette smoke, UV rays in sunlight, pollutants from car exhausts or any of the thousands of molecules and radiation sources that we're exposed to every day – must impact on our genes and cause a change in expression.

The majority of non-infectious diseases that afflict most people take a long time to develop, and then remain as a problem for many years if there is no cure available. The stimuli from the environment could theoretically be acting on the genes all the time in the cells that are acting abnormally, leading to disease. But this seems unlikely, especially because most of the chronic diseases probably involve the interaction of multiple stimuli with multiple genes. It's hard to imagine that all these stimuli would be present for decades at a time. The alternative is that there is a mechanism that keeps the disease-associated cells in an abnormal state, i.e. expressing genes inappropriately.

In the absence of any substantial evidence for a role for somatic mutation, epigenetics seems like a strong candidate for this mechanism. This would allow the genes in one twin to stay mis-regulated, ultimately leading to a disease. We're only at the beginning of the investigation but some evidence has started accumulating that suggests this may indeed be the case.

One of the most straightforward experiments conceptually, is to analyse if chromatin modification patterns (the epigenome) change as MZ twins get older. In the simplest case, we wouldn't even need to investigate this in the context of disease. We could start by testing a much simpler hypothesis – that genetically identical individuals become epigenetically non-identical as they age. If this hypothesis is correct, this would support the idea that MZ twins can vary from each other at the epigenetic level. This in turn would strengthen our confidence in moving forwards to examining the role of epigenetic changes in disease.

In 2005, a large collaborative group headed by Professor Manel Esteller, then at the Spanish National Cancer Centre in Madrid, published a paper in which they examined this issue[1]. They made some interesting discoveries. If they examined chromatin from infant MZ twin pairs, they couldn't see much difference in the levels of DNA methylation or of histone acetylation between the two twins. When they looked at pairs of MZ twins who were much older, such as in their fifties, there was a lot of variation within the pair for the amount of DNA methylation or histone acetylation. This seemed to be particularly true of twins that had lived apart for a long time.

The results from this study were consistent with a model where genetically identical twins start out epigenetically very similar, and then diverge as they get older. The older MZ twins who had led separate and different lives for the longest would be expected to be the ones who had encountered the greatest differences in their environments. The finding that these were precisely the twin pairs who were most different epigenetically was consistent with the idea that the epigenome (the overall pattern of epigenetic modifications on the genome) reflects environmental differences.

Children who eat breakfast are statistically more likely to do well at school than children who skip breakfast. This doesn't necessarily mean that learning can be improved by a bowl of cornflakes. It may simply be that children who eat breakfast are more likely to be children whose parents make an effort to get them to school every day, on time, and help them with their studies. Similarly, Professor Esteller's data are correlative. They show there is a relationship between the ages of twins and how different they are epigenetically, but they don't prove that age has *caused* the change in the epigenome. But at least the hypothesis can remain in play.

A team led by Dr Jeffrey Craig in 2010 at the Royal Children's Hospital in Melbourne also examined DNA methylation in identical and fraternal twin pairs[2]. They investigated a few relatively small regions of the genome in greater detail than in Manel Esteller's earlier paper. Using samples just from newborn twin pairs, they showed that there was a substantial amount of difference between the DNA methylation patterns of fraternal twins. This isn't unexpected, since fraternal twins are genetically non-identical and we expect different individuals to have different epigenomes. Interestingly, though, they also found that even the MZ twins differed in their DNA methylation patterns, suggesting identical twins begin to diverge epigenetically during development in the uterus. Combining the information from the two papers, and from additional studies, we can conclude that even genetically identical individuals are epigenetically distinct by the time of birth, and these epigenetic differences become more pronounced with age and exposure to different environments.

Of mice and men (and women)

These data are consistent with a model where epigenetic changes could account for at least some of the reasons why MZ twins aren't phenotypically identical, but there's still a lot of supposition involved. That's because for many purposes humans are a quite hopeless experimental system. If we want to be able to assess the role of epigenetics in the problem of why genetically identical individuals are phenotypically different from one another, we would like to be able to do the following:

1. Analyse hundreds of identical individuals, not just pairs of them;
2. Manipulate their environments, in completely controlled ways;
3. Transfer embryos or babies from one mother to another, to investigate the effects of early nurture;
4. Take all sorts of samples from the different tissues of the body, at lots of different time points;
5. Control who mates with whom;
6. Carry out studies on four or five generations of genetically identical individuals.

Needless to say, this isn't feasible for humans.

This is why experimental animals have been so useful in epigenetics. They allow scientists to address really complex questions, whilst controlling the environment as much as possible. The data that are generated in these animal studies produce insights from which we can then try to infer things about humans.

The match may not be perfect, but we can unravel a surprising amount of fundamental biology this way. Various comparative studies have shown that many systems have stayed broadly the same in different organisms over almost inconceivably long periods. The epigenetic machinery of yeast and humans, for example, share more similarities than differences and yet the common ancestor for the two species lies about one billion years in the past[3]. So, epigenetic processes are clearly fairly fundamental things, and using model systems can at least point us in a helpful direction for understanding the human condition.

In terms of the specific question we've been looking at in this chapter – why genetically identical twins often don't seem to be identical – the animal that has been most useful is our close mammalian relative, the mouse. The mouse and human lineages separated a mere 75 million or so years ago[4]. 99 per cent of the genes found in mice can also be detected in humans, although they aren't generally absolutely identical between the two species.

Scientists have been able to create strains of mice in which all the individuals are genetically identical to each other. These have been incredibly useful for investigating the roles of non-genetic factors in creating variation between individuals. Instead of just two genetically identical individuals, it's possible to create hundreds, or thousands. The way this is done would have made even the Ptolemy dynasty of ancient Egypt blush. Scientists mate a pair of mice who are brother and sister. Then they mate a brother and sister from the resulting litter. They then mate a brother and sister from their litter and so on. When this is repeated for over twenty generations of brother-sister matings, all the genetic variation gets bred out, throughout the genome. All mice of the same sex from the strain are genetically identical. In a refinement of this, scientists can take these genetically identical mice and introduce just one change into their DNA. They may use such genetic engineering to create mice which are identical except for just one region of DNA that the experimenters are most interested in.

A mouse of a different colour

The most useful mouse model for exploring how epigenetic changes can lead to phenotypic differences between genetically identical individuals is called the *agouti* mouse. Normal mice have hair which is banded in colour. The hair is black at the tip, yellow in the middle and black again at the base. A gene called *agouti* is essential for creating the yellow bit in the middle, and is switched on as part of a normal cyclical mechanism in mice.

There is a mutated version of the *agouti* gene (called *a*) which never switches on. Mice that only have the *a*, mutant version of *agouti* have hair which is completely black. There is also a particular mutant mouse strain called A^{vy}, which stands for *agouti viable yellow*. In A^{vy} mice, the *agouti* gene is switched on permanently and the hair is yellow through its entire length. Mice have two copies of the *agouti* gene, one inherited from the mother and one from the father. The A^{vy} version of the gene is dominant to the *a* version, which means that if one copy of the gene is A^{vy} and one is *a*, the A^{vy} will 'overrule' *a* and the hairs will be yellow throughout their length. This is all summarised in Figure 5.2.

Scientists created a strain of mice that contained one copy of A^{vy} and one copy of *a* in every cell. The nomenclature for this is A^{vy}/a. Since A^{vy} is dominant to *a*, you would predict that the

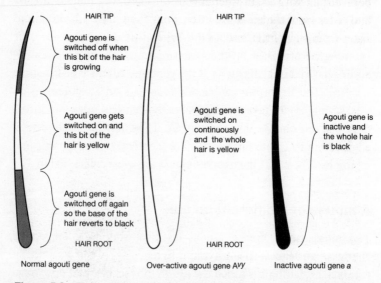

HAIR TIP

Agouti gene is switched off when this bit of the hair is growing

Agouti gene gets switched on and this bit of the hair is yellow

Agouti gene is switched off again so the base of the hair reverts to black

HAIR ROOT

Normal agouti gene

HAIR TIP

Agouti gene is switched on continuously and the whole hair is yellow

HAIR ROOT

Over-active agouti gene A^{vy}

Agouti gene is inactive and the whole hair is black

Inactive agouti gene *a*

Figure 5.2   Hair colour in mice is affected by the expression of the *agouti* gene. In normal mice, the agouti protein is expressed cyclically, leading to the characteristic brindled pattern of mouse fur. Disruption of this cyclical pattern of expression can lead to hairs which are either yellow or black throughout their length.

mice would have completely yellow hair. Since all the mice in the strain are genetically identical, you would expect that they would all look the same. But they don't. Some have the very yellow fur, some the classic mouse appearance caused by the banded fur, and some are all shades in-between, as shown in Figure 5.3.

This is really odd, since the mice are all genetically exactly the same. All the mice have the same DNA code. We could argue that perhaps the differences in coat colour are due to environment, but laboratory conditions are so standardised that this seems unlikely. It's also unlikely because these differences can be seen in mice from the same litter. We would expect mice from a single litter to have very similar environments indeed.

Of course, the beauty of working with mice, and especially with highly inbred strains, is that it's relatively easy to perform detailed genetic and epigenetic studies, especially when we already have a reasonable idea of where to look. In this case, the region to examine was the *agouti* gene.

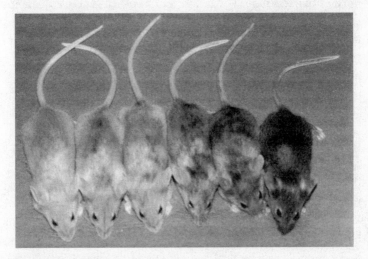

Figure 5.3   Genetically identical mice showing the extent to which fur colour can vary, depending on expression of the agouti protein. Photo reproduced with the kind permission of Professor Emma Whitelaw.

Mouse geneticists knew how the yellow phenotype was caused in A^{vy} yellow mice. A piece of DNA had been inserted in the mouse chromosome just before the *agouti* gene. This piece of DNA is called a retrotransposon, and it's one of those DNA sequences that doesn't code for a protein. Instead, it codes for an abnormal piece of RNA. Expression of this RNA messes up the usual control of the downstream *agouti* gene and keeps the gene switched on continuously. This is why the hairs on the A^{vy} mice are yellow rather than banded.

That still doesn't answer the question of why genetically identical A^{vy}/a mice had variable coat colour. The answer to this has been shown to be due to epigenetics. In some A^{vy}/a mice the CpG sequences in the retrotransposon DNA have become very heavily methylated. As we saw in the previous chapter, DNA methylation of this kind switches off gene expression. The retrotransposon no longer expressed the abnormal RNA that messed up transcription from the *agouti* gene. These mice were the ones with fairly normal banded mouse coat colour. On other genetically identical A^{vy} mice, the retrotransposon was unmethylated. It produced its troublesome RNA which messed up the transcription from the *agouti* gene so that it was switched on continuously and the mice were yellow. Mice with in-between levels of retrotransposon methylation had in-between levels of yellow fur. This model is shown in Figure 5.4.

Here, DNA methylation is effectively working like a dimmer switch. When the retrotransposon is unmethylated, it shines to its fullest extent, producing lots of the abnormal RNA. The more the retrotranposon is methylated, the more its expression gets turned down.

The *agouti* mouse has provided a quite clear-cut example of how epigenetic modification, in this case DNA methylation, can make genetically identical individuals look phenotypically different. However, there is always the fear that *agouti* is a special case, and maybe this is a very uncommon mechanism. This is

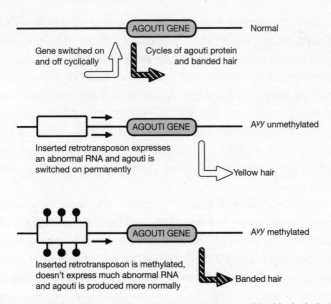

Figure 5.4   Variations in DNA methylation (represented by black circles) influence expression of a retrotransposon. The variation in expression of the retrotransposon in turn affects expression of the agouti gene, leading to coat colour variability between genetically identical animals.

particularly of concern because it's proved very difficult to find an *agouti* gene in humans – it seems to be in that 1 per cent of genes we don't share with our mouse neighbours.

There is another interesting condition found in mice, in which the tail is kinked. This is called Axin-fused and it also demonstrates extreme variability between genetically identical individuals. This has been shown to be another example where the variability is caused by differing levels of DNA methylation in a retrotransposon in different animals, just like the *agouti* mouse.

This is encouraging as it suggests this mechanism isn't a one off, but kinked tails still don't really represent a phenotype that is of much concern to the average human. But there's something we can all get on board with: body weight. Genetically identical mice don't all have the same body weight.

No matter how tightly scientists control the environment for the mice, and especially their access to food, identical mice from inbred mouse strains don't all have exactly the same body weight. Experiments carried out over many years have shown that only about 20–30 per cent of the variation in body weights can be attributed to the post-natal environment. This leaves the question of what causes the other 70–80 per cent of variation in body weight[5]. Since it isn't being caused by genetics (all the mice are identical) or by the environment, there has to be another source for the variation.

In 2010, Professor Emma Whitelaw, the terrifically enthusiastic and intensely rigorous mouse geneticist working at the Queensland Institute of Medical Research, published a fascinating paper. She used an inbred strain of mice and then used genetic engineering to create subsets of animals which were genetically identical to the starting stock, except that they only expressed half of the normal levels of a particular epigenetic protein. She performed the genetic engineering independently in a number of mice, so that she could create separate groups of animals, each of which was mutated in a different gene coding for epigenetic proteins.

When Professor Whitelaw analysed the body weights of large numbers of the normal or mutated mice, an interesting effect appeared. In a group of normal inbred mice, most of the animals had relatively similar body weights, within the ranges found in many other studies. In the mice with low levels of a certain epigenetic protein, there was a lot more variability in the body weights within the group. Further experiments published in the same paper assessed the effects of the decreased expression of these epigenetic proteins. Their decreased expression was linked to changes in expression levels of selected genes involved in metabolism[6], and increased variability in that expression. In other words, the epigenetic proteins were exerting some control over the expression of other genes, just as we might expect.

Emma Whitelaw tested a number of epigenetic proteins in her system, and found that only a few of them caused the increased variation in body weight. One of the proteins that had this effect was Dnmt3a. This is one of the enzymes that transfers methyl groups to DNA, to switch genes off. The other epigenetic protein that caused increased variability in body weight was called Trim28. Trim28 forms a complex with a number of other epigenetic proteins which together add specific modifications to histones. These modifications down-regulate expression of genes near the modified histones and are known as repressive histone modifications or marks. Regions of the genome that have lots of repressive marks on their histones tend to become methylated on their DNA, so the Trim28 may be important for creating the right environment for DNA methylation.

These experiments suggested that certain epigenetic proteins act as a kind of dampening field. 'Naked' DNA is rather prone to being switched on somewhat randomly, and the overall effect is like having a lot of background chatter in our cells. This is called transcriptional noise. The epigenetic proteins act to turn down the volume of this random chat. They do this by covering the histones with modifications that reduce the genes' expression. It's likely that different epigenetic proteins are important for suppressing different genes in some tissues rather than in others.

It's clear that this suppression isn't total. If it were, then all inbred mice would be identical in every aspect of their phenotype and we know this isn't the case. There is variation in body weight even in the inbred strains, it's just that there's even more variation in the mice with the depressed levels of the epigenetic proteins.

This sophisticated balancing act, in which epigenetic proteins dampen down transcriptional noise but don't entirely repress gene expression, is a cellular compromise. It leaves cells with enough flexibility of gene expression to be able to respond to new signals – be these hormones or nutrients, pollutants or sunlight – but without the genes being constantly ready to fire up just for the heck of it. Epigenetics allows cells to perform the difficult compromise between becoming (and remaining) different cell types with a variety of functions, and not being so locked into a single pattern of gene expression that they become incapable of responding to changes in their environment.

Something that is becoming increasingly clear is that early development is a key period when this control of transcriptional noise first becomes established. After all, very little of the variation in body weight in the original inbred strains could be attributed to the post-natal environment (just 20–30 per cent). Interest is increasing all the time in the role of a phenomenon called developmental programming, whereby events during foetal development can impact on the whole of adult life, and it is increasingly recognised that epigenetic mechanisms are what underlie a major proportion of this programming.

Such a model is entirely consistent with Emma Whitelaw's work on the effects of decreased levels of Dnmt3a or Trim28 in her mouse studies. The body weight effects were apparent when the mice were just three weeks old. This model is also consistent with the fact that decreased levels of Dnmt3a resulted in the increased variability in body weight, but decreased levels of the related enzyme Dnmt1 had no effect in Emma Whitelaw's experiments. Dnmt3a can add methyl groups to totally unmethylated DNA regions, which means it is responsible for establishing the correct DNA methylation patterns in cells. Dnmt1 is the protein that maintains pre-established methylation patterns on DNA. It seems that the most important feature for dampening down gene expression variability (at least as far as body weight is concerned) is establishing the correct DNA methylation patterns in the first place.

The Dutch Hunger Winter

Scientists and policy-makers have recognised for many years the importance of good maternal health and nutrition during pregnancy, to increase the chances that babies will be born at a healthy weight and so be more likely to thrive physically. In more recent years, it's become increasingly clear that if a mother is malnourished during pregnancy, her child may be at increased risk of ill-health, not just during the immediate post-birth infancy, but for decades. We've only recently begun to realise that this is at least in part due to molecular epigenetic effects, which result in impaired developmental programming and life-long defects in gene expression and cellular function.

As already highlighted, there are extremely powerful ethical and logistical reasons why humans are a difficult species to use experimentally. Tragically, historical events, terrible at the time, conspire to create human scientific study groups by accident. One of the most famous examples of this is the Dutch Hunger Winter, which was mentioned in the Introduction.

This was a period of terrible hardship and near-starvation during the Nazi fuel and food blockade of the Netherlands in the last winter of the Second World War. Twenty-two thousand people died and the desperate population ate anything they could find, from tulip bulbs to animal blood. The dreadful privations of the population created a remarkable scientific study population. The Dutch survivors were a well-defined group of individuals all of whom suffered just one period of malnutrition, all of them at exactly the same time.

One of the first aspects to be studied was the effect of the famine on the birthweights of children who had been in the womb during the famine. If a mother was well-fed around the time of conception and malnourished only for the last few months of the pregnancy, her baby was likely to be born small. If, on the other hand, the mother suffered malnutrition for the first three months of the pregnancy only (because the baby was conceived towards the end of this terrible episode), but then was well-fed, she was likely to have a baby with normal body weight. The foetus 'caught up' in body weight, because foetuses do most of their growing in the last few months of pregnancy.

But here's the thing – epidemiologists were able to study these groups of babies for decades and what they found was really surprising. The babies who were born small stayed small all their lives, with lower obesity rates than the general population. Even more unexpectedly, the adults whose mothers had been malnourished only early in their pregnancy had higher obesity rates than normal. Recent reports have shown a greater incidence of other health problems as well, including certain aspects of mental health. If mothers suffered severe malnutrition during the early stages of pregnancy, their children were more likely than usual to develop schizophrenia. This has been found not just in the Dutch Hunger Winter cohort but also in the survivors of the monstrous Great Chinese Famine of 1958 to 1961, in which millions starved to death as a result of Mao Tse Tung's policies.

Even though these individuals had seemed perfectly healthy at birth, something that had happened during their development in the womb affected them for decades afterwards. And it wasn't just the fact that something had happened that mattered, it was *when* it happened. Events that take place in the first three months of development, a stage when the foetus is really very small, can affect an individual for the rest of their life.

This is completely consistent with the model of developmental programming, and the epigenetic basis to this. In the early stages of pregnancy, where different cell types are developing, epigenetic proteins are probably vital for stabilising gene expression patterns. But remember that our cells contain thousands of genes, spread over billions of base-pairs, and we have hundreds of epigenetic proteins. Even in normal development there are likely to be slight variations in the expression of some of these proteins, and the precise effects that they have at specific chromosomal regions. A little bit more DNA methylation here, a little bit less there.

The epigenetic machinery reinforces and then maintains particular patterns of modifications, thus creating the levels of gene expression. Consequently, these initial small fluctuations in histone and DNA modifications may eventually become 'set' and get transmitted to daughter cells, or be maintained in long-lived cells such as neurons, that can last for decades. Because the epigenome gets 'stuck', so too may the patterns of gene expression in certain chromosomal regions. In the short term the consequences of this may be relatively minor. But over decades all these mild abnormalities in gene expression, resulting from a slightly inappropriate set of chromatin modifications, may lead to a gradually increasing functional impairment. Clinically, we don't recognise this until it passes some invisible threshold and the patient begins to show symptoms.

The epigenetic variation that occurs in developmental programming is at heart a predominantly random process, normally referred to as 'stochastic'. This stochastic process may account for a significant amount of the variability that develops between the MZ twins who opened this chapter. Random fluctuations in epigenetic modifications during early development lead to non-identical patterns of gene expression. These become epigenetically set and exaggerated over the years, until eventually the genetically identical twins become phenotypically different, sometimes in the most dramatic of ways. Such a random process, caused by individually minor fluctuations in the expression of epigenetic genes during early development also provides a very good model for understanding how genetically identical A^{vy}/a mice can end up with different coat colours. This can be caused by randomly varying levels of DNA methylation of the A^{vy} retrotransposon.

Such stochastic changes in the epigenome are the likely reason why even in a totally inbred mouse strain, kept under completely standardised conditions, there is variation in body weight. But once a big environmental stimulus is introduced in addition to this stochastic variation, the variability can become even more pronounced.

A major metabolic disturbance during early pregnancy, such as the dramatically decreased availability of food during the Dutch Hunger Winter, would significantly alter the epigenetic processes occurring in the foetal cells. The cells would change metabolically, in an attempt to keep the foetus growing as healthily as possible despite the decreased nutrient supply. The cells would change their gene expression to compensate for the poor nutrition, and the patterns of expression would be set for the future because of epigenetic modifications to the genes. It's probably no surprise that it was the children whose mothers had been malnourished during the very early stages of pregnancy, when developmental programming is at its peak, who went on to be at higher risk of adult obesity. Their cells had become epigenetically programmed to make the most of limited food supply. This programming remained in place even when the environmental condition that had prompted it – famine – was long over.

Recent studies examining DNA methylation patterns in the Dutch Hunger Winter survivors have shown changes at key genes involved in metabolism. Although a correlation like this doesn't prove cause-and-effect, the data are consistent with under-nutrition during the early developmental period changing the epigenomic profile of key metabolic genes[7].

It's important to recognise that even in the Dutch Hunger Winter cohort, the effects that we see are not all-or-nothing. Not every individual whose mother had been malnourished early in pregnancy became obese. When scientists studied the population they found an increased *likelihood* of adult obesity. This is again consistent with a model where random epigenetic variability, the genotypes of the individuals and early environmental events, and the responses of the genes and cells to the environment combine in one great big complicated – and as yet not easily decipherable – equation.

Severe malnutrition is not the only factor that has effects on a foetus that can last a lifetime. Excessive alcohol consumption during pregnancy is a leading preventable cause of birth defects and mental retardation (foetal alcohol syndrome) in the Western world[8]. Emma Whitelaw used the *agouti* mouse to investigate if alcohol can alter the epigenetic modifications in a mouse model of foetal alcohol syndrome. As we have seen, expression of the A^{vy} gene is epigenetically controlled via DNA methylation of a retrotransposon. Any stimulus that alters DNA methylation of the retrotransposon would change expression of the A^{vy} gene. This would affect the colour of the fur. In this model, fur colour becomes a 'read-out' that indicates changes in epigenetic modifications.

Pregnant mice were given free access to alcohol. The coat colour in the pups from the alcohol-drinking mothers was compared with the coat colour of the pups from pregnant mice that didn't have access to booze. The distribution of coat colours was different between the two groups. So were the levels of DNA methylation of the retrotransposon, as predicted. This showed that the alcohol had led to a change in the epigenetic modifications in the mice. Disruption of epigenetic developmental programming may lead to at least some of the debilitating and lifelong symptoms of foetal alcohol syndrome in children of mothers who over-use alcohol during pregnancy.

Bisphenol A is a compound used in the manufacture of polycarbonate plastics. Feeding bisphenol A to *agouti* mice results in a change in the distribution of coat colour, suggesting this chemical has effects on developmental programming through epigenetic mechanisms. In 2011 the European Union outlawed bisphenol A in drinking bottles for babies.

Early programming may also be one of the reasons that it's been very difficult to identify the environmental effects that lead to some chronic human conditions. If we study pairs of MZ twins who are discordant for a specific phenotype, for example multiple sclerosis, it may be well nigh impossible to identify an environmental cause. It may simply be that one of the pair was exceptionally unlucky in the random epigenetic fluctuations that established certain key patterns of gene expression early in life. Scientists are now mapping the distribution of epigenetic changes in concordant and discordant MZ twins for a number of disorders, to try to identify histone or DNA modifications that correlate with the presence or absence of disease.

Children conceived during famines and mice with yellow coats have each clearly taught us remarkable things about early development, and the importance of epigenetics in this process. Oddly enough, these two disparate groups have one other thing to teach us. At the very beginning of the 19th century, Jean-Baptiste Lamarck published his most famous work, *Philosophie Zoologique*. He hypothesised that acquired characteristics can be transmitted from one generation to the next, and that this drives evolution. As an example, a short-necked giraffe-like animal that elongated its neck by constant stretching would pass on a longer neck to its offspring. This theory has been generally dismissed and in most cases it is simply wrong. But the Dutch Hunger Winter cohort and the yellow mice have shown us that startlingly, the heretical Lamarckian model of inheritance can, just sometimes, be right on the money, as we are about to see.

Chapter 6

The Sins of the Fathers

*For I, the Lord your God, am a jealous God, punishing
the children for the sins of the fathers to the third and
fourth generation of those who hate me*
Exodus, Chapter 20, Verse 5

The *Just So* stories published by Rudyard Kipling at the very
beginning of the 20th century are an imaginative set of tales about
origins. Some of the most famous are those about the phenotypes
of animals – *How the Leopard Got his Spots, The Beginning of
the Armadillos, How the Camel Got his Hump*. They are written
purely as entertaining fantasies but scientifically they hark back to
a century earlier and Lamarck's theory of evolution through the
inheritance of acquired characteristics. Kipling's stories describe
how one animal acquired a physical characteristic – the elephant's
long trunk, for example – and the implication is that all the offspring inherited that characteristic, and hence all elephants now
have long trunks.

Kipling was having fun with his stories, whereas Lamarck
was trying to develop a scientific theory. Like any good scientist,
he tried to collect data relevant to this hypothesis. In one of the
most famous examples of this, Lamarck recorded that the sons
of blacksmiths (a very physical trade) tended to have larger arm
muscles than the sons of weavers (a much less physical occupation). Lamarck interpreted this as the blacksmiths' sons inherit-
ing the acquired phenotype of large muscles from their fathers.

Our modern interpretation is different. We recognise that
a man whose genes tended to endow him with the ability to
develop large muscles would be at an advantage in a trade such

as blacksmithing. This occupation would attract those who were genetically best suited to it. Our interpretation would also encompass the likelihood that the blacksmith's sons may have inherited this genetic tendency towards chunky biceps. Finally, we would acknowledge that at the time that Lamarck was writing, children were used routinely as additional members of a family workforce. The children of a blacksmith were more likely than those of a weaver to be performing relatively heavy manual labour from an early age and hence would be likely to develop larger arm muscles as a response to their environment, just as we all do when we pump iron.

It would be a mistake to look back on Lamarck and only mock. We no longer accept most of his ideas scientifically, but we should acknowledge that he was making a genuine attempt to address important questions. Inevitably, and quite rightly, Lamarck has been overshadowed by Charles Darwin, the true colossus of 19th century biology – actually, probably the colossus of biology generally. Darwin's model of the evolution of species via natural selection has been the single most powerful conceptual framework in biological sciences. Its power became even greater once married to Mendel's work on inheritance and our molecular understanding of DNA as the raw material of inheritance.

If we wanted to summarise a century and a half of evolutionary theory in one paragraph we might say:

Random variation in genes creates phenotypic variation in individuals. Some individuals will survive better than others in a particular environment, and these individuals are likely to have more offspring. These offspring may inherit the same advantageous genetic variation as their parent, so they too will have increased breeding success. Eventually, over a huge number of generations, separate species will evolve.

The raw material for random variation is mutation of the DNA sequence of the individual; his or her genome. Mutation rates are generally very low, and so it takes a long time for advantageous mutations to develop and to spread through a population. This is especially the case if each mutation only gives an individual a slight advantage over its competitors in a particular environment.

This is where the Lamarckian model of acquired characteristics really falls over, relative to Darwinian models. An acquired change in phenotype would somehow have to 'feed-back' onto the DNA script and change it really dramatically, so that the acquired characteristic could be transmitted in the space of just one generation, from parent to child. But there's very little evidence that this happens, except occasionally in response to chemicals or irradiation which damage DNA (mutagens), causing a change in the actual base-pair sequence. Even these mutagens only affect the genome at a relatively small percentage of base-pairs and in a random pattern, so these still can't drive inheritance of acquired characteristics in any meaningful way.

The overwhelming body of data argues against Lamarckian inheritance, so there's very little reason for individual scientists to work on this experimentally. This isn't surprising. After all, if you are a scientist interested in the Solar System, you could choose to investigate the hypothesis that at least some parts of the Moon are made of cheese. But to do so would mean that you wilfully ignored the large body of evidence already present against this – hardly a rational approach.

There's also possibly a cultural reason that scientists have shied away from experimental investigations of the inheritance of acquired characteristics. One of the most notorious cases of scientific fraud is that of Paul Kammerer, who worked in Austria in the first half of the 20th century. He claimed that he had demonstrated the inheritance of acquired characteristics in a species called the midwife toad.

Kammerer reported that when he changed the conditions in which the toads bred, they developed 'useful' adaptations. These adaptations were structures on their forelimbs called nuptial pads, which were black in colour. Unfortunately, very few of the specimens were retained or stored well, and when a rival scientist examined a specimen he found that India ink had been injected into the pad. Kammerer denied all knowledge of the contamination and killed himself shortly afterwards. This scandal tainted an already controversial field[1].

One of the statements in our potted history of evolutionary theory was the following, 'An acquired change in phenotype would somehow have to 'feed-back' onto the DNA script and change it really dramatically so that the acquired characteristic could be transmitted in the space of just one generation, from parent to child.'

It's certainly hard to imagine how an environmental influence on the cells of an individual could act at a specific gene to change the base-pair sequence. But it's all too obvious that epigenetic modifications – be these DNA methylation or alterations to the histone proteins – do indeed occur at specific genes in response to the environmental influences on a cell. The response to hormonal signalling that was mentioned in an earlier chapter was an example of this. Typically, a hormone like oestrogen will bind to a receptor on a cell from, for example, the breast. The oestrogen and the receptor stay together and move into the nucleus of the cell. They bind to specific motifs in DNA – A, C, G and T bases in a particular sequence – which are found at the promoters of certain genes. This helps to switch on the genes. When it binds to these motifs, the oestrogen receptor also attracts various epigenetic enzymes. These alter the histone modifications, removing marks that repress gene expression and putting on marks that tend to switch genes on. In this way, the environment, acting via hormones, can change the epigenetic pattern at specific genes.

These epigenetic modifications don't change the sequence of a gene, but they do alter how the gene is expressed. This is, after all, the whole basis of developmental programming for later disease. We know that epigenetic modifications can be transmitted from a parent cell to a daughter cell, as this is why there are no teeth in your eyeballs. If a similar mechanism transmitted an environmentally-induced epigenetic modification from an individual to their offspring, we would have a mechanism for a sort of Lamarckian inheritance. An epigenetic (as opposed to genetic) change would be passed down from parent to child.

Heresy and the Dutch Hunger Winter

It's all very well to think about how this could happen, but really we need to know if acquired characteristics can actually be inherited in this way. Not *how* does it happen, but the more basic question of *does* it happen? Remarkably, there appear to be some specific situations where this is indeed taking place. This doesn't mean that Darwinian/Mendelian models are wrong, it just means that, as always, the world of biology is more complicated than we imagined.

The scientific literature on this contains some confusing terminology. Some early papers refer to epigenetic transmission of an acquired trait but don't seem to have any evidence of DNA methylation changes, or histone alterations. This isn't sloppiness on the part of the authors. It's because of the different ways in which the word epigenetics has been used. In the early papers the phrase 'epigenetic transmission' refers to inheritance that cannot be explained by genetics. In these cases, the word epigenetic is being used to describe the phenomenon, not the molecular mechanism. To try to keep everything a little clearer, we'll use the phrase 'transgenerational inheritance' to describe the phenomenon of transmission of an acquired characteristic and only use 'epigenetics' to describe molecular events.

Some of the strongest evidence for transgenerational inheritance in humans comes from the survivors of the Dutch Hunger Winter. Because the Netherlands has such excellent medical infrastructure, and high standards of patient data collection and retention, it has been possible for epidemiologists to follow the survivors of the period of famine for many years. Significantly, they were able to monitor not just the people who had been alive in the Dutch Hunger Winter, but also their children and their grandchildren.

This monitoring identified an extraordinary effect. As we have already seen, when pregnant women suffered malnutrition during the first three months of the pregnancy, their babies were born with normal weight, but in adulthood were at higher risk of obesity and other disorders. Bizarrely, when women from this set of babies became mothers themselves, their first born child tended to be heavier than in control groups[2,3]. This is shown in Figure 6.1, where the relative sizes of the babies have been exaggerated for clarity, and where we've given the women arbitrary Dutch names.

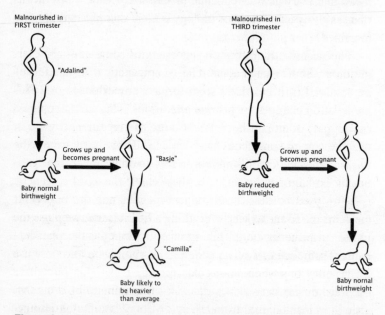

Figure 6.1   The effects of malnutrition across two generations of children and grandchildren of women who were pregnant during the Dutch Hunger Winter. The timing of the malnutrition in pregnancy was critical for the subsequent effects on body weight.

The effects on the birth weight of baby Camilla shown at the bottom left are really odd. When Camilla was developing, her mother Basje was presumably healthy. The only period of malnutrition that Basje had suffered was twenty or more years earlier, when she was going through her own first stages of development in the womb. Yet it seems that this has an effect on her own child, even though Camilla was never exposed to a period of malnutrition during early development.

This seems like a good example of transgenerational (Lamarckian) inheritance, but has it has been caused by an epigenetic mechanism? Did an epigenetic change (altered DNA methylation and/or variations in histone modifications) that had occurred in Basje as a result of malnutrition during her first twelve weeks of development in the womb get passed on via the nucleus of her egg to her own child? Maybe, but we shouldn't ignore that there are other potential explanations.

For example, there could be an unidentified effect of the early malnutrition which means that when pregnant, Basje will pass more nutrients than normal across the placenta to her foetus. This would still create a transgenerational effect – Camilla's increased size – but it wouldn't be caused by Basje passing on an epigenetic modification to Camilla. It would be caused by the conditions in the womb when Camilla was developing and growing (the intra-uterine environment).

It's also important to remember that a human egg is large. It contains a nucleus which is relatively small in volume compared to the surrounding cytoplasm. Imagine a grape inside a satsuma to gain some idea of relative sizes. The cytoplasm carries out a lot of functions when an egg gets fertilised. Perhaps something occurred during early developmental programming in Basje that ultimately resulted in the cytoplasm of her eggs containing something unusual. That might sound unlikely but egg production in female mammals is actually initiated early in their own embryonic development. The earliest stages of zygote development rely to a large extent on the cytoplasm from the egg. An abnormality in the cytoplasm could stimulate an unusual growth pattern in the foetus. This again would result in transgenerational inheritance but not through the direct transmission of an epigenetic modification.

So we can see that there are various mechanisms that could explain the inheritance patterns seen through the maternal line in the Dutch Hunger Winter survivors. It would help us to understand if epigenetics plays a role in acquired inheritance if we could study a less complicated human situation. Ideally, this would be a scenario where we don't have to worry about the effects of the intra-uterine environment, or the cytoplasm of the egg.

Let's hear it for fathers. Because men don't get pregnant, they can't contribute to the developmental environment of the foetus. Males also don't contribute much cytoplasm to the zygote. Sperm are very small and are almost all nucleus – they look like little bullets with tails attached. So if we see transgenerational inheritance from father to child, it isn't likely to be caused by intra-uterine or cytoplasmic effects. Under these circumstances, an epigenetic mechanism would be an attractive candidate for explaining transgenerational inheritance of an acquired characteristic.

Greedy fellows in Sweden

Some data suggesting that male transgenerational inheritance can occur in humans comes from another historical study. There is a geographically isolated region in Northern Sweden called Överkalix. In the late 19th and early 20th centuries there were periods of terrible food shortages (caused by failed harvests, military actions and transport inadequacies), interspersed with periods of great plenty. Scientists have studied the mortality patterns for descendants of people who were alive during these periods. In particular, they analysed food intake during a stage in childhood known as the slow growth period (SGP). All other factors being equal, children grow slowest in the years leading up to puberty. This is a completely normal phenomenon, seen in most populations.

Using historical records, the researchers deduced that if food was scarce during a father's SGP, his son was at decreased risk of dying through cardiovascular disease (such as stroke, high blood pressure or coronary artery disease). If, on the other hand, a man had access to a surfeit of food during the SGP, his grandsons were at increased risk of dying as a consequence of diabetic ill-nesses[4]. Just like Camilla in the Dutch Hunger Winter example, the sons and grandsons had an altered phenotype (a change in the risk of death through cardiovascular disease or diabetes) in response to an environmental challenge they themselves had never experienced.

These data can't be a result of the intra-uterine environ-ment nor of cytoplasmic effects, for the reasons outlined earlier. Therefore, it seems reasonable to hypothesise that the transgen-erational consequences of food availability in the grandparental generation were mediated via epigenetics. These data are par-ticularly striking when you consider that the original nutritional effect happened when the boys were pre-pubescent and so had not even begun to produce sperm. Even so, they were able to pass an effect on to their sons and grandsons.

However, there are some caveats around this work on transgen-erational inheritance through the male line. In particular, there are risks involved in relying on old death records, and extrapolat-ing backwards through historical data. Additionally, some of the effects that were observed were not terribly large. This is frequently a problem when working with human populations, along with all the other issues we have already discussed, such as our genetic variability and the impossibility of controlling environment in any major way. There is always the risk that we draw inappropri-ate conclusions from our data, rather as we believe Lamarck did with his studies on the families of blacksmiths.

The heretical mouse

Is there an alternative way of investigating transgenerational inheritance? If this phenomenon also occurs in other species, it would give us a lot more confidence that these effects are real. This is because experiments in model systems can be designed to test specific hypotheses, rather than just using the datasets that nature (or history) provides.

This is where we come back to the *agouti* mouse. Emma Whitelaw's work showed that the variable coat colour in the *agouti* mouse was due to an epigenetic mechanism, specifically DNA methylation of a retrotransposon in the *agouti* gene. Mice of different colour all had the same DNA sequence, but a different degree of epigenetic modification at the retrotransposon.

Professor Whitelaw decided to investigate if the coat colour could be inherited. If it could, it would show that it's not only DNA that gets transmitted from parent to offspring, but also epigenetic modifications to the genome. This would provide a potential mechanism for the transgenerational inheritance of acquired characteristics.

When Emma Whitelaw allowed female *agouti* mice to breed, she found the effect that is shown in Figure 6.2. For convenience, the picture only shows the offspring who inherited the A^{vy} retrotransposon from their mother, as this is the effect we are interested in.

If the mother had an unmethylated A^{vy} gene, and hence had yellow fur, all her offspring also had either yellow fur, or slightly mottled fur. She never had offspring who developed the very dark fur associated with the methylation of the retrotransposon.

By contrast, if the mother's A^{vy} gene was heavily methylated, resulting in her having dark fur, some of her offspring also had dark fur. If both grandmother and mother had dark fur, then the effect was even more pronounced. About a third of the final offspring had dark fur, compared with the one in five shown in Figure 6.2.

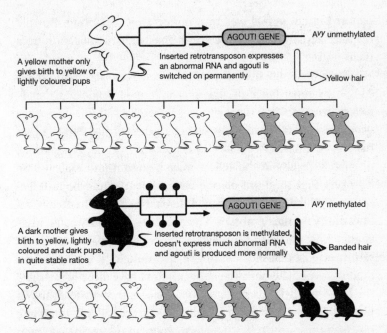

Figure 6.2 The coat colour of genetically identical female mice influences the coat colour of their offspring. Yellow female mice, in whom the agouti gene is expressed continuously, due to low levels of DNA methylation of the regulatory retrotransposon, never give birth to dark pups. The epigenetically – rather than genetically – determined characteristics of the mother influence her offspring.

Because Emma Whitelaw was working on inbred mice, she was able to perform this experiment multiple times and generate hundreds of genetically identical offspring. This was important, as the more data points we have in an experiment, the more we can rely on the findings. Statistical tests showed that the phenotypic differences between the genetically identical groups were highly significant. In other words, it was very unlikely that the effects occurred by chance[5].

The results from these experiments showed that an epigenetically-mediated effect (the DNA methylation-dependent coat pattern) in an animal was transmitted to its offspring. But did the mice actually inherit directly an epigenetic modification from their mother?

There was a possibility that the effects seen were not directly caused by inheritance of the epigenetic modification at the A^{vy} retrotransposon, but through some other mechanism. When the *agouti* gene is switched on too much, it doesn't just cause yellow fur. *Agouti* also mis-regulates the expression of other genes, which ultimately results in the yellow mice being fat and diabetic. So it's likely that the intra-uterine environment would be different between yellow and dark pregnant females, with different nutrient availability for their embryos. The nutrient availability could itself change how particular epigenetic marks get deposited at the A^{vy} retrotransposon in the offspring. This would look like epigenetic inheritance, but actually the pups wouldn't have directly inherited the DNA methylation pattern from their mother. Instead, they'd just have gone through a similar developmental programming process in response to nutrient availability in the uterus.

Indeed, at the time of Emma Whitelaw's work, scientists already knew that diet could influence coat colour in *agouti* mice. When pregnant *agouti* mice are fed a diet rich in the chemicals that can supply methyl groups to the cells (methyl donors), the ratios of the differently coloured pups changes[6]. This is presumably because the cells are able to use more methyl groups, and deposit more methylation on their DNA, hence shutting down the abnormal expression of *agouti*. This meant that the Whitelaw group had to be really careful to control for the effect of intra-uterine nutrition in their experiments.

In one of those experiments that simply aren't possible in humans, they transferred fertilised eggs obtained from yellow mothers and implanted them into dark females, and vice versa. In every case, the distribution of coat patterns in the offspring was the same as was to be expected from the egg donor, i.e. the biological mother, rather than the surrogate. This showed unequivocally that it wasn't the intra-uterine environment that controlled the coat patterning. By using complex breeding schemes, they also demonstrated that the inheritance of the coat pattern was not due to the cytoplasm in the egg. Taken together, the most straightforward interpretation of these data is that epigenetic inheritance has taken place. In other words, an epigenetic modification (probably DNA methylation) was transferred along with the genetic code.

This transfer of the phenotype from one generation to the next wasn't perfect – not all the offspring looked exactly the same as their mother. This implies that the DNA methylation that controls the expression of the *agouti* phenotype wasn't entirely stable down the generations. This is quite analogous to the effects we see in suspected cases of human transgenerational inheritance, such as the Dutch Hunger Winter. If we look at a large enough number of people in our study group we can detect differences in birth weight between various groups, but we can't make absolute predictions about a single individual.

There is also an unusual gender-specific phenomenon in the *agouti* strain. Although coat pattern showed a clear transgenerational effect when it was passed on from mother to pup, no such effect was seen when a male mouse passed on the A^{vy} retrotransposon to his offspring. It didn't matter if a male mouse was yellow, lightly mottled or dark. When he fathered a litter, there were likely to be all the different patterns of colour in his offspring.

But there are other examples of epigenetic inheritance transmitted from both males and females. The kinked tail phenotype in mice, which is caused by variable methylation of a retrotransposon in the $Axin^{Fu}$ (Axin fused) gene, can be transmitted by either the mother or the father[7]. This makes it unlikely that transgenerational inheritance of this characteristic is due to intra-uterine or cytoplasmic influences, because fathers don't really contribute much to these. It's far more likely that there is the transmission of an epigenetic modification at the $Axin^{Fu}$ gene from either parent to offspring.

These model systems have been really useful in demonstrating that transgenerational inheritance of a non-genetic phenotype does actually occur, and that this takes place via epigenetic modifications. This is truly revolutionary. It confirms that for some very specific situations Lamarckian inheritance is taking place, and we have a handle on the molecular mechanism behind it. But the *agouti* and kinked tail phenotypes in mice both rely on the presence of specific retrotransposons in the genome. Are these special cases, or is there a more general effect in play? Once again, we return to something that has a bit more immediate relevance for us all. Food.

The epigenetics of obesity

As we all know, an obesity epidemic is developing. It's spreading worldwide, although it's advancing at a particularly fast rate in the more industrialised societies. The frankly terrifying graph in Figure 6.3 displays the UK figures for 2007[8], showing that about two out of every three adults is overweight (body mass index of

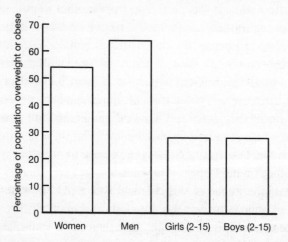

Figure 6.3 The percentage of the UK population that was overweight or obese in 2007.

25 or over) or obese (body mass index of 30 or over). The situation is even worse in the USA. Obesity is associated with a wide range of health problems including cardiovascular disease and type 2 diabetes. Obese individuals over the age of 40 will die, on average, 6 to 7 years earlier than non-obese people[9].

The data from the Dutch Hunger Winter and other famines support the idea that poor nutrition during pregnancy has effects on offspring, and that these consequences can be transmitted to subsequent generations as well. In other words, poor nutrition can have epigenetic effects on later generations. The data from the Överkalix cohort, although more difficult to interpret, suggested that excess consumption at key points in a boy's life can have adverse consequences for later generations. Is it possible that the obesity epidemic in the human population will have knock-on effects for children and grandchildren? As we don't really want to wait 40 years to work this out, scientists are again turning to animal models to try to gain some useful insights.

The first animal data suggested that nutrition might not have much effect transgenerationally. The change in coat pattern of pups when pregnant *agouti* mice were given diets high in methyl donors didn't transmit to the next generation[10]. But perhaps this is too specialised a model. In 2010, two papers were published that should at least give us pause for thought. They were published in two of the best journals in the world – *Nature* and *Cell*. In both cases, the researchers overfed male animals and then monitored the effects on their offspring. By restricting their experiments to males, they didn't need to worry about the intra-uterine and cytoplasmic complications that cause such (metaphorical) headaches if studying females.

One of the studies used a breed of rat called Sprague-Dawley. This is an albino rat, with a chilled-out temperament that makes it easy to keep and handle. In the experiments male Sprague-Dawleys were given a high-fat diet, and allowed to mate with females who had been fed an ordinary diet. The over-fed males were overweight (hardly a surprise), had a high percentage of fat to muscle and had many of the symptoms found in type 2 diabetes in humans. Offspring were normal weight but they too had the diabetes-type abnormalities[11]. Many of the genes that control metabolism and how mammals burn fuel were mis-regulated in these offspring. For reasons that aren't understood, it was particularly the daughters that showed this effect.

A completely independent group studied the effects of diet in an inbred mouse strain. Male mice were fed a diet that was abnormally low in protein. The diet had an increased percentage of sugar to make up for this. The males were mated to females on a normal diet. The researchers examined the expression of genes in the liver (the body's major organ when it comes to metabolism) in three-week-old pups from these matings. Analysing large numbers of mouse pups, they found that the regulation of many of the genes involved in metabolism was abnormal in the offspring of the males that had been fed the modified diet[12]. They also found changes in the epigenetic modifications in the livers of these pups.

So, both these studies show us that, at least in rodents, a father's diet can directly influence the epigenetic modifications, gene expression and health of his offspring. And not because of environment – this isn't like the human example of a child getting fat because their Dad only ever feeds them super-sized portions of burgers and chips. It's a direct effect and it occurred so frequently in the rats and mice that it can't have been due to diet-induced mutations, they just don't happen at that sort of rate. So the most likely explanation is that diet induces epigenetic effects that can be transmitted from father to child. Although the data are quite preliminary, the results from the mouse study in particular support this.

If you look at all the data in its entirety – from humans to rodents, from famine to feast – a quite worrying pattern emerges. Maybe the old saw of 'we are what we eat' doesn't go far enough. Maybe we're also what our parents ate and what their parents ate before them.

This might make us wonder if there is any point following advice on healthy living. If we are all victims of epigenetic determinism, this would suggest that our dice have already been rolled, and we are just at the mercy of our ancestors' methylation patterns. But this is far too simplistic a model. Overwhelming amounts of data show that the health advice issued by government agencies and charities – eating a healthy diet rich in fruit and vegetables, getting off the sofa, not smoking – is completely sound. We are complex organisms, and our health and life expectancy are influenced by our genome, our epigenome and our environment. But remember that even in the inbred *agouti* mice, kept under standardised conditions, researchers couldn't predict exactly how yellow or how fat an individual mouse in a newborn litter would become. Why not do everything that we can to improve our chances of a healthy and long life? And if we are planning to have children, don't we want to do whatever we can to nudge them that bit closer to good health?

There will always be things we can't control, of course. One of the best-documented examples of an environmental factor that has epigenetic consequences, lasting at least four generations, is an environmental toxin. Vinclozolin is a fungicide, which tends to be used particularly frequently in the wine industry. If it gets into mammals it is converted into a compound that binds to the androgen receptor. This is the receptor that binds testosterone, the male hormone that is vital for sexual development, sperm production and a host of other effects in males. When vinclozolin binds to the androgen receptor, it prevents testosterone from transmitting its usual signals to the cells, and so blocks the normal effects of the hormone.

If vinclozolin is given to pregnant rats at the time when the testes are developing in the embryos, the male offspring are born with testicular defects and have reduced fertility. The same effect is found for the next three generations[13]. About 90 per cent of the male rats are affected, which is far too high a percentage to be caused by classic DNA mutation. Even the highest known rates of mutation, at particularly sensitive regions of the genome, are at least ten-fold less frequent than this. In these rat experiments, only one generation was exposed to vinclozolin, yet the effect lasted for at least four generations, so this is another example of Lamarckian inheritance. Given the male transmission pattern, it is likely this is another example of an epigenetic inheritance mechanism. A follow-on publication from the same research group has identified regions of the genome where vinclozolin treatment leads to unusual DNA methylation patterns[14].

The rats in the studies described above were treated with high doses of vinclozolin. These were much larger than humans are believed to encounter in the environment. Nonetheless, effects such as these are one of the reasons why some authorities are beginning to investigate if artificial hormones and hormone disrupters in the environment (from excretion of chemicals present in the contraceptive pill, to certain pesticides) have the potential to cause subtle, but potentially transgenerational effects in the human population.

Chapter 7

The Generations Game

The animals went in two by two, hurrah! Hurrah!
Traditional song

Sometimes, the best science starts with the simplest of questions. The question may seem so obvious that almost nobody thinks to ask it, let alone answer it. We just don't challenge things that seem completely self-evident. Yet occasionally, when someone stands up and asks, 'How does that happen?', we all realise that a phenomenon that seems too obvious to mention, is actually a complete mystery. This is true of one of the most fundamental aspects of human biology, one we almost never think about.

When mammals (including humans) reproduce, why does this require a male and a female parent?

In sexual reproduction the small, very energetic sperm swim like crazy to get to the large, relatively sedentary egg. When a winning sperm penetrates the egg, the nuclei from the two cells fuse to create the zygote that divides to form every cell in the body. Sperm and eggs are referred to as gametes. When gametes are produced in the mammalian body, each gamete receives only half the normal number of chromosomes. This means they only have 23 chromosomes, one of each pair. This is known as a haploid genome. When the two nuclei fuse after a sperm has penetrated the egg, the chromosome number is restored to that of all ordinary cells (46) and the genome is called diploid. It's important that the egg and the sperm are both haploid, otherwise each generation would end up with twice as many chromosomes as its parents.

We could hypothesise that the reason why mammals all have a mother and father is because that's what we need to introduce two haploid genomes to one another, to create a new cell with a full complement of chromosomes. Certainly it's true that this is what normally happens but this model would also imply that the only reason why biologically we need a parent of each sex is because of a delivery system.

Conrad Waddington's grandson

In 2010 Professor Robert Edwards received the Nobel Prize in Physiology or Medicine for his pioneering work in the field of in vitro fertilisation, which led to the so-called test tube babies. In this work, eggs were removed from a woman's body, fertilised in the laboratory, and re-implanted back into the uterus. In vitro fertilisation was hugely challenging, and Professor Edwards' success in human reproduction was built on years of painstaking work in mice.

This mouse work laid the foundation for a remarkable series of experiments, which demonstrated there's a lot more to mammalian reproduction than just a delivery system. The major force in this field is Professor Azim Surani, from Cambridge University, who started his scientific career by obtaining his PhD under the supervision of Robert Edwards. Since Professor Edwards received his early research training in Conrad Waddington's lab, we can think of Azim Surani as Conrad Waddington's intellectual grandson.

Azim Surani is another of those UK academics who carries his prestige very lightly, despite his status. He is a Fellow of the Royal Society and a Commander of the British Empire, and has been awarded the prestigious Gabor Medal and Royal Society Royal Medal. Like John Gurdon and Adrian Bird, he continues to break new ground in a research area that he pioneered over a quarter of a century ago.

Starting in the mid 1980s, Azim Surani carried out a programme of experiments which showed unequivocally that mammalian reproduction is much more than a matter of a delivery system. We don't just need a biological mother and a biological father because that's how two haploid genomes fuse to form one diploid nucleus. It actually matters enormously that half of our DNA comes from our mother and half from our father.

Figure 7.1 shows what a just-fertilised egg looks like, before the two genomes meet. It's simplified and exaggerated, but it will serve our purpose. The haploid nuclei from the egg and the sperm are called pro-nuclei.

We can see that the female pronucleus is much bigger than the male one. This is very important experimentally, as it means that we can tell the different pronuclei apart. Because we can tell them apart, scientists can transfer a pronucleus from one cell to another, and be certain about which one they transferred. They know if they transferred a pronucleus that came from the father's sperm (male pronucleus) or from the mother's egg (female pronucleus).

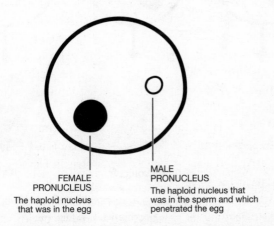

FEMALE
PRONUCLEUS
The haploid nucleus
that was in the egg

MALE
PRONUCLEUS
The haploid nucleus that
was in the sperm and which
penetrated the egg

Figure 7.1 The mammalian egg just after it has been penetrated by a sperm, but before the two haploid (half the normal chromosome number) pronuclei have fused. Note the disparity in size between the pronucleus that came from the egg, and the one that originated from the sperm.

Many years ago Professor Gurdon used tiny micropipettes to transfer the nuclei from the body cells of toads into toad eggs. Azim Surani used a refinement of this technology to transfer pronuclei between different fertilised eggs from mice. The manipulated fertilised eggs were then implanted into female mice and allowed to develop.

In a slew of papers, mainly published between the years of 1984 and 1987, Professor Surani demonstrated that it's essential to have a male and a female pronucleus in order to create new living mice. This is shown graphically in Figure 7.2.

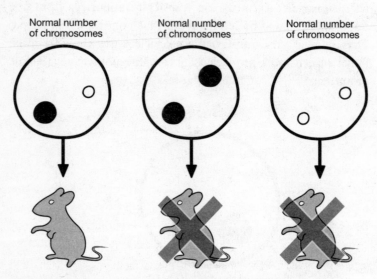

Figure 7.2   A summary of the outcomes from the early work of Azim Surani. The pronucleus was removed from a mouse egg. This donor egg was then injected with two haploid pronuclei and the resulting diploid egg was implanted into a mouse surrogate mother. Live mice resulted only from the eggs which had been reconstituted with one male and one female pronucleus. Embryos from eggs reconstituted with either two male or two female pronuclei failed to develop properly and the embryos died during development.

To control for the effects of different DNA genomes, the researchers used inbred mouse strains. This ensured that the three types of fertilised eggs shown in the diagram were genetically identical. Yet despite being genetically identical, a series of experiments from Azim Surani and his colleagues[1,2,3], along with other work from the laboratories of Davor Solter[4] and Bruce Cattanach[5] were conclusive. If the fertilised egg contained only two female pronuclei, or two male ones, no live mice were ever born. You needed a pronucleus of each sex.

This is an absolutely remarkable finding. In all three scenarios shown in the diagram, the zygote ends up with exactly the same amount of genetic material. Each zygote has a diploid genome (two copies of every chromosome). If the only factor that was important in the creation of a new individual was the *amount* of DNA, then all three types of fertilised eggs should have developed to form new individuals.

Quantity isn't everything

This led to a revolutionary concept – the maternal and paternal genomes may deliver the same DNA but they are not functionally equivalent. It's not enough just to have the correct amount of the correct sequence of DNA. We have to inherit some from our mother and some from our father. Somehow, our genes 'remember' who they come from. They will only function properly if they come from the 'correct' parent. Just having the right number of copies of each gene, doesn't fulfil the requirements for normal development and healthy life.

We know that this isn't some strange effect that only applies to mice, because of a naturally occurring human condition. In about one in 1500 human pregnancies, for example, there is a placenta in the uterus but there is no foetus. The placenta is abnormal, covered in fluid-filled, grape-like lumps. This structure is called a hydatidiform mole, and in some Asian populations the frequency of these molar pregnancies can be as high as 1 in 200. The apparently pregnant women gain weight, often more quickly than in a normal pregnancy and they also suffer morning sickness, often to a quite extreme degree. The rapidly-growing placental structures produce abnormally high levels of a hormone which is thought to be responsible for the symptoms of nausea in pregnancy.

In countries with good healthcare infrastructure, the hydatidiform mole is normally detected at the first ultrasound scan, and then an abortion-type procedure is carried out by a medical team. If not detected, the mole will usually abort spontaneously at around four or five months post-fertilisation. Early detection of these moles is important as they can form potentially dangerous tumours if they aren't removed.

These moles are formed if an egg which has somehow lost its nucleus is fertilised. In about 80 per cent of hydatidiform molar pregnancies, an empty egg is fertilised by a single sperm, and the haploid sperm genome is copied to create a diploid genome. In about 20 per cent of cases the empty egg is fertilised simultaneously by two sperm. In both cases the fertilised egg has the correct number of chromosomes (46), but all the DNA came from the father. Because of this, no foetus develops. Just like the experimental mice, human development requires chromosomes from both the mother and the father.

This human condition and the experiments in mice are impossible to reconcile with a model based only on the DNA code, where DNA is a naked molecule, which carries information only in its sequence of A, C, G and T base-pairs. DNA alone isn't carrying all the necessary information for the creation of new life. Something else must be required in addition to the genetic information. Something epigenetic.

Eggs and sperm are highly specialised cells – they are at the bottom of one of Waddington's troughs. The egg and the sperm will never be anything other than an egg and a sperm. Unless they fuse. Once they fuse, these two highly specialised cells form one cell which is so unspecialised it is totipotent and gives rise to every cell in the human body, and the placenta. This is the zygote, at the very top of Waddington's epigenetic landscape. As this zygote divides, the cells become more and more specialised, forming all the tissues of our bodies. Some of these tissues ultimately give rise to eggs or sperm (depending on our sex, obviously) and the whole cycle is ready to start again. There's effectively a never-ending circle in developmental biology.

The chromosomes in the pro-nuclei of sperm and eggs carry large numbers of epigenetic modifications. This is part of what keeps these gametes behaving as gametes, and not turning into other cell types. But these gametes can't be passing on their epigenetic patterns, because if they did the fertilised zygote would be some sort of half-egg, half-sperm hybrid when it clearly isn't this at all. It's a completely different totipotent cell that will give rise to an entirely new individual. Somehow the modifications on eggs and sperm get changed to a different set of modifications, to drive the fertilised egg into a different cell state, at a different position in Waddington's epigenetic landscape. This is part of normal development.

Re-installing the operating system

Almost immediately after the sperm has penetrated the egg, something very dramatic happens to it. Pretty much all the methylation on the male pronucleus DNA (i.e. from the sperm) gets stripped off, incredibly quickly. The same thing happens to the DNA on the female pronucleus, albeit a lot more slowly. This means that a lot of the epigenetic memory gets wiped off the genome. This is vital for putting the zygote at the top of Waddington's epigenetic landscape. The zygote starts dividing and soon creates the blastocyst – the golf ball inside the tennis ball from Chapter 2. The cells in the golf ball – the inner cell mass, or ICM – are the pluripotent cells, the ones that give rise to embryonic stem cells in the laboratory.

The cells of the ICM soon differentiate and start giving rise to the different cell types of our bodies. This happens through very tightly regulated expression of a few key genes. One specific protein, for example OCT4, switches on another set of genes, which results in a further cascade of gene expression, and so on. We have met *OCT4* before – it is the most critical of all the genes that Professor Yamanaka used to reprogram somatic cells. These cascades of gene expression are associated with epigenetic modification of the genome, changing the DNA and histone marks so that certain genes stay switched on or get switched off appropriately. Here's the sequence of epigenetic events in very early development:

1. The male and female pronuclei (from the sperm and the egg respectively) are carrying epigenetic modifications;
2. The epigenetic modifications get taken off (in the immediate post-fertilisation zygote);
3. New epigenetic modifications get put on (as the cells begin to specialise).

This is a bit of a simplification. It's certainly true that researchers can detect huge swathes of DNA demethylation during stage 2 from this list. However, it's actually more complicated than this, particularly in respect of histone modifications. Whilst some histone modifications are being removed, others are becoming established. At the same time as the repressive DNA methylation is removed, certain histone marks which repress gene expression are also erased. Other histone modifications which increase gene expression may take their place. It's therefore too naïve to refer to the epigenetic changes as just being about putting on or taking off epigenetic modifications. In reality, the epigenome is being reprogrammed.

Reprogramming is what John Gurdon demonstrated in his ground-breaking work when he transferred the nuclei from adult toads into toad eggs. It's what happened when Keith Campbell and Ian Wilmut cloned Dolly the Sheep by putting the nucleus from a mammary gland cell into an egg. It's what Yamanaka achieved when he treated somatic cells with four key genes, all of which code for proteins highly expressed naturally during this reprogramming phase.

The egg is a wonderful thing, honed through hundreds of millions of years of evolution to be extraordinarily effective at generating vast quantities of epigenetic change, across billions of base-pairs. None of the artificial means of reprogramming cells comes close to the natural process in terms of speed or efficiency. But the egg probably doesn't quite do everything unaided. At the very least, the pattern of epigenetic modifications in sperm is one that allows the male pronucleus to be reprogrammed relatively easily. The sperm epigenome is primed to be reprogrammed[6].

Unfortunately, these priming chromatin modifications (and many other features of the sperm nucleus), are missing if an *adult* nucleus is reprogrammed by transferring it into a fertilised egg. That's also true when an adult nucleus is reprogrammed by treating it with the four Yamanaka factors to create iPS cells. In both these circumstances, it's a real challenge to completely reset the epigenome of the adult nucleus. It's just too big a task.

This is probably why so many cloned animals have abnormalities and shortened lifespans. The defects that are seen in these cloned animals are another demonstration that if early epigenetic modifications go wrong, they may stay wrong for life. The abnormal epigenetic modification patterns result in permanently inappropriate gene expression, and long-term ill-health.

All this reprogramming of the genome in normal early development changes the epigenome of the gametes and creates the new epigenome of the zygote. This ensures that the gene expression patterns of eggs and sperm are replaced by the gene expression patterns of the zygote and the subsequent developmental stages. But this reprogramming also has another effect. Cells can accumulate inappropriate or abnormal epigenetic modifications at various genes. These disrupt normal gene expression and can even contribute to disease, as we shall see later in this book. The reprogramming of the egg and the sperm prevent them from passing on from parent to offspring any inappropriate epigenetic modifications they have accumulated. Not so much wiping the slate clean, more like re-installing the operating system.

Making the switch

But this creates a paradox. Azim Surani's experiments showed that the male and female pro-nuclei aren't functionally equivalent; we need one of each to create a new mammal. This is known as a parent-of-origin effect, because it essentially shows that there are ways for a zygote and its daughter cells to distinguish between chromosomes from the mother and father. This isn't a genetic effect, it is an epigenetic one, and so there must be some epigenetic modifications that do get transmitted from one generation to the next.

In 1987 the Surani lab published one of the first papers to give an insight into this mechanism. They hypothesised that parent-of-origin effects could be caused by DNA methylation. At that time, this was really the only chromatin modification that had been identified, so it was an excellent place to start. The researchers created genetically modified mice. These mice contained an extra piece of DNA that could get inserted randomly anywhere in the genome. The DNA sequence of this extra bit wasn't particularly important to the experimenters. What was important was that they could easily measure how much DNA methylation was present on this sequence, and whether the amount of methylation was transmitted faithfully from parent to offspring.

Azim Surani and his colleagues examined seven lines of mice with this random insertion. In one of the seven lines, something very odd happened. When a mother passed on the inserted DNA, it was always heavily methylated in her offspring. But when a male mouse passed it on to his offspring, the mouse pups always ended up with low methylation of this foreign DNA. Figure 7.3 demonstrates this.

Black represents the methylated inserted DNA, whereas white represents unmethylated DNA. Fathers always give their offspring white, unmethylated DNA whereas mothers always give their offspring black, methylated DNA. In other words, the methylation in the offspring is dependent on the *sex* of the parent who passed the inserted DNA onto them. It's not dependent on what the methylation was like in the parent. For example, a 'black' male will always have offspring with 'white' DNA.

What this paper from Azim Surani[7], and another published at the same time[8], demonstrated was that when mammals create eggs

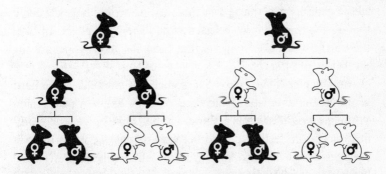

Figure 7.3   Mice generated in which a particular foreign piece of DNA was either methylated or not methylated. Black represents methylated DNA, and white represents unmethylated. When a mother passed on this foreign DNA, the DNA was always heavily methylated (black) in her offspring, regardless of whether she herself had been 'black' or 'white'. The opposite was found for males, whose offspring always had unmethylated 'white' DNA. This was the first experimental demonstration that some regions of the genome can be marked to indicate if they were inherited via the maternal or the paternal line.

and sperm, they somehow manage to barcode the DNA in these cells. It's as if the chromosomes carry little flags. The chromosomes in sperm carry little flags that say, 'I'm from Dad' and the chromosomes in eggs carry little flags that say, 'I'm from Mum'. DNA methylation is the fabric that these flags are made from.

The description that is used for this is imprinting – the chromosomes have been imprinted with information about which parent they came from originally. Imprinting and parent-of-origin effects are something we will explore in more detail in the next chapter.

What was happening to the foreign DNA in the experiments, which kept changing its methylation status as it was transmitted from parent to offspring? It had, quite by chance, got inserted into a region of the mouse DNA that carried one of these flags. As a consequence, the foreign DNA also started getting DNA methylation flags stuck to it when it was passed down the generations.

The fact that only one of seven mouse lines showed this effect suggested that not all of the genome carries these flags. If the whole genome was marked in this way, we would have expected that all the lines that were tested would show the effect. In fact, the one in seven rate suggests that these flagged regions are the exception, not the rule.

In Chapter 6 we saw that sometimes animals do inherit acquired characteristics from their parents. The work of Emma Whitelaw, amongst others, shows us that some epigenetic modifications do indeed get passed between parent and offspring, via the sperm and the egg. This type of inheritance is pretty rare, but it does strengthen our belief that there must be some epigenetic modifications that are special. They don't get replaced when the egg and sperm fuse to form the zygote. So, although the vast majority of the mammalian genome does get reset when the egg and the sperm fuse, a small percentage of it is immune from this reprogramming.

The epigenetics arms race

Only 2 per cent of our genome codes for proteins. A massive 42 per cent is composed of retrotransposons. These are very odd sequences of DNA, which probably originated from viruses in our evolutionary past. Some retrotransposons are transcribed to produce RNA and this can affect the expression of neighbouring genes. This can have serious consequences for cells. If it drives up expression of genes that cause cells to proliferate too aggressively, for example, this may nudge cells towards becoming cancerous.

There's a constant arms race in evolution, and mechanisms have evolved in our cells to control the activity of these types of retrotransposons. One of the major mechanisms that cells use is epigenetic. The retrotransposon DNA gets methylated by the cell, turning off retrotransposon RNA expression. This prevents the RNA disrupting expression of neighbouring genes. One particular class, known as IAP retrotransposons, seems to be a particular target of this control mechanism.

During reprogramming in the early zygote, the methylation is removed from most of our DNA. But IAP retrotransposons are an exception to this. The reprogramming machinery has evolved to skip these rogue elements and leave the DNA methylation marks on them. This keeps the retrotransposons in an epigenetically repressed state. This has probably evolved as a mechanism to reduce the risk that potentially dangerous IAP retrotransposons will get accidentally re-activated.

This is relevant because the two best-studied examples of transgenerational inheritance of non-genetic features are the *agouti* mouse and the *Axin^{Fu}* mouse, which we met in the previous chapter. The phenotypes in both these models are a consequence of the methylation levels of an IAP retrotransposon upstream of a gene. The DNA methylation levels in the parent get passed on to the offspring, and so does the phenotype caused by the expression levels of the retrotransposon[9].

We met other examples of transgenerational inheritance of acquired characteristics in Chapter 6, including the effects of nutrition on subsequent generations, and the transgenerational effects of environmental pollutants such as vinclozolin. Researchers are exploring the hypothesis that these environmental stimuli create epigenetic changes in the chromatin of the gametes. These alterations are probably in regions that are protected from reprogramming during early development after the egg and sperm fuse.

Like John Gurdon, Azim Surani has continued to work highly productively in a field that he pioneered. His work has been focused on how and why eggs and sperm barcode their DNA so that a molecular memory is passed on to the next generation. A large amount of Azim Surani's initial pioneering work was dependent on manipulating mammalian nuclei by using tiny pipettes to transfer them between cells. Technically, this is a refined version of the methods that John Gurdon used so successfully fifteen years earlier. It's oddly pleasing to consider that Professor Surani is now based at the research institute in Cambridge that is named after Professor Gurdon, and that they frequently bump into each other in the corridors and the coffee room.

Chapter 8

The Battle of the Sexes

Nobody will ever win the Battle of the Sexes. There's just too much fraternising with the enemy.
Henry Kissinger

The laboratory stick insect *Carausius morosus* is a very popular pet. As long as it has a few privet leaves to munch on it will be perfectly content, and after a few months it will begin to lay eggs. In due course, these will hatch into perfect little baby stick insects, looking just like miniature versions of the adults. If one of these baby stick insects is removed as soon as it is born, and kept in a tank on its own, then it too will lay eggs which will hatch into little stick insects in their turn. This is despite the fact that it has never mated.

Stick insects frequently reproduce this way. They are using a mechanism known as parthenogenesis, from the Greek for 'virgin birth'. Females lay fertile eggs without ever mating with a male, and perfectly healthy little stick insects emerge from these eggs. These insects have evolved with special mechanisms to ensure that the offspring have the correct number of chromosomes. But these chromosomes all came from the mother.

This is very different from mice and humans, as we saw in the last chapter. For us and our rodent relatives, the only way to generate live young is by having DNA from both a mother and a father. It's tempting to speculate that stick insects are highly unusual but they're not. We mammals are the exceptions. Insects, fish, amphibians, reptiles and even birds all have a few species that can reproduce parthenogenetically. It's we mammals who can't. It's our class in the animal kingdom which is the odd one out, so it makes sense to ask why this is the case. We can begin by looking at the features which are found only in mammals. Well, we have hair, and we have three bones in our middle ear. Neither of these characteristics is found in the other classes, but it seems unlikely these are the key features that have led us to abandon virgin birth. For this issue there is a much more important characteristic.

The most primitive examples of mammals are the small number of creatures like the duck-billed platypus and the echidna, which lay eggs. After them, in terms of reproductive complexity, are the marsupials such as the kangaroo and the Tasmanian devil, which give birth to very under-developed young. The young of these species go through most of their developmental stages outside the mother's body, in her pouch. The pouch is a glorified pocket on the outside of the body.

By far the greatest numbers of our class are called placental (or eutherian) mammals. Humans, tigers, mice, blue whales – we all nourish our young in the same way. Our offspring undergo a really long developmental phase inside the mother, in the uterus. During this developmental stage, the young get their nourishment via the placenta. This large, pancake-shaped structure acts as an interface between the blood system of the foetus and the blood system of the mother. Blood doesn't actually flow from one to the other. Instead the two blood systems pass so closely to one another that nutrients such as sugars, vitamins, minerals and amino acids can pass from the mother to the foetus. Oxygen also passes from the mother's blood to the foetal blood supply. In exchange, the foetus gets rid of waste gases and other potentially harmful toxins by passing them back into the mother's circulation.

It's a very impressive system, and allows mammals to nurture their young for long periods during early development. A new placenta is created in each pregnancy and the code for its production isn't carried by the mother. It's all coded for by the foetus. Think back yet again to our model of the early blastocyst in Chapter 2. All the cells of the blastocyst are descendants of the fertilised single-cell zygote. The cells that will ultimately become the placenta are the tennis ball cells on the outside of the blastocyst. In fact, one of the earliest decisions that cells make as they begin to roll down Waddington's epigenetic landscape is whether they are turning into future placental cells, or future body cells.

We can't escape our (evolutionary) past

While the placenta is a great method for nourishing a foetus, the system has 'issues'. To use business or political speech, there's a conflict of interest, because in evolutionary terms, our bodies are faced with a dilemma.

This is the evolutionary imperative for the male mammal, phrased anthropomorphically:

> *This pregnant female is carrying my genes in the form of this foetus. I may never mate with her again. I want my foetus to get as big as possible so that it has the greatest chance of passing on my genes.*

For the female mammal, the evolutionary imperative is rather different:

> *I want this foetus to survive and pass on my genes. But I don't want it to be at the cost of draining me so much that I never reproduce again. I want more than this one chance to pass on my genes.*

This battle of the sexes in mammals has reached an evolutionary Mexican stand-off. A series of checks and balances ensures that neither the maternal nor the paternal genome gets the upper hand. We can get a better understanding of how this works if we look once again at the experiments of Azim Surani, Davor Sobel and Bruce Cattanach. These are the scientists who created the mouse zygotes that contained only paternal DNA or only maternal DNA.

After they had created these test tube zygotes, the scientists implanted them into the uterus of mice. None of the labs ever generated living mice from these zygotes. However, the zygotes did develop for a while in the womb, but very abnormally. The abnormal development was quite different, depending on whether all the chromosomes had come from the mother or the father.

In both cases the few embryos that did form were small and retarded in growth. Where all the chromosomes had come from the mother, the placental tissues were very underdeveloped[1]. If all the chromosomes came from the father, the embryo was even more retarded but there was much better production of the placental tissues[2]. Scientists created embryos from a mix of these cells – cells which had only maternally inherited or paternally inherited chromosomes. These embryos still couldn't develop all the way to birth. When examined, the researchers found that all the tissues in the embryo were from the maternal-only cells whereas the cells of the placental tissues were the paternal-only type[3].

All these data suggested that something in the male chromosomes pushes the developmental programme in favour of the placenta, whereas a maternally-derived genome has less of a drive towards the placenta, and more towards the embryo itself. How is this consistent with the conflict or evolutionary imperative laid out earlier in this chapter? Well, the placenta is the portal for taking nutrients out of the mother and transferring them into the foetus. The paternally-derived chromosomes promote placental development, and thereby create mechanisms for diverting as much nutrition as possible from the mother's bloodstream. The maternal chromosomes act in the opposite way, and a finely poised stalemate develops in normal pregnancies.

One obvious question is whether all the chromosomes are important for these effects. Bruce Cattanach used complex genetic experiments on mice to investigate this. The mice contained chromosomes that had been rearranged in different ways. The simplest way to explain this is that each mouse had the right amount of chromosomes, but they'd been 'stuck together' in unusual ways. He was able to create mice which had precise abnormalities in the inheritance of their chromosomes. For example, he could create mice which inherited both copies of a specific chromosome from just one parent.

The first experiments he reported were using mouse chromosome 11. For all the other pairs of chromosomes, the mice inherited one of each pair maternally, and one paternally. But for chromosome 11, Bruce Cattanach created mice that had inherited two copies from their mother and none from their father, or vice versa. Figure 8.1 represents the results[4].

Once again this is consistent with the idea that there are factors in the paternal chromosomes that push towards development of larger offspring. Factors in the maternal chromosomes either act in the 'opposite direction' or are broadly neutral.

As we explored in the last chapter, these factors are epigenetic, not genetic. In the example above, let's assume that the parents came from the same inbred mouse strain, so were genetically identical. If you sequenced both copies of chromosome 11 in any

| Source of chromosome 11 | Both copies from mother. Mouse smaller than normal. | One copy from each parent. Normal mouse. | Both copies from father. Mouse larger than normal. |

Figure 8.1   Bruce Cattanach created genetically modified mice, in which he could control how they inherited a particular region of chromosome 11. The middle mouse inherited one copy from each parent. Mice which inherited both copies from their mother were smaller than this normal mouse. In contrast, mice which inherited both copies from their father were larger than normal.

of the three types of offspring, they would be exactly the same. They would contain the same millions of A, C, G and T base-pairs, in the same order. But the two copies of chromosome 11 do clearly behave differently at a functional level, as shown by the different sizes of the different types of mice. Therefore there must be epigenetic differences between the maternal and paternal copies of chromosome 11.

Sex discrimination

Because the two copies of the chromosome behave differently depending on their parent-of-origin, chromosome 11 is known as an imprinted chromosome. It has been imprinted with information about its origins. As our understanding of genetics has improved we've realised that only certain stretches of chromosome 11 are imprinted. There are large regions where it doesn't matter at all which parent donated which chromosome, and the regions from the two parents are functionally equivalent. There are also entire chromosomes that are not imprinted.

So far, we've described imprinting in mainly phenomenological terms. Imprinted regions are stretches of the genome where we can detect parent-of-origin effects in offspring. But how do these regions carry this effect? In imprinted regions, certain genes are switched on or switched off, depending on how they were inherited. In the chromosome 11 example above, genes associated with placental growth are switched on and are very active in the copy of the chromosome inherited from the father. This carries risks of nutrient depletion for the mother who is carrying the foetus, and a compensatory mechanism has evolved. The copies of these same genes on the maternal chromosome tend to be switched off, and this limits the placental growth. Alternatively, there may be other genes that counterbalance the effects of the paternal genes, and these counter-balancing genes may be expressed mainly from the maternal chromosome.

Major strides have been made in understanding the molecular biology of these effects. For example, later researchers worked on a region on chromosome 7 in mice. There is a gene in this region called *insulin-like growth factor 2* (*Igf2*). The Igf2 protein promotes embryonic growth, and is normally expressed only from the paternally-derived copy of chromosome 7. Experimenters introduced a mutation into this gene, which stopped the gene coding for a functional Igf2 protein. They studied the effects of the mutation on offspring. When the mutation was passed on from the mother, the young mice looked the same as any other mice. This is because the *Igf2* gene is normally switched off on the maternal chromosome anyway, and so it didn't matter that the maternal gene was mutated. But when the mutant *Igf2* gene was passed down from father to offspring, the mice in the litter were much smaller than usual. This was because the one copy of the *Igf2* gene that they 'relied on' for strong foetal growth had been switched off by the mutation[5].

There is a gene on mouse chromosome 17 called *Igf2r*. The protein encoded by this gene 'mops up' Igf2 protein and stops it acting as a growth promoter. The *Igf2r* gene is also imprinted. Because Igf2r protein has the 'opposite' effect to Igf2 in terms of foetal growth, it probably isn't surprising to learn that the *Igf2r* gene is usually expressed from the maternal copy of chromosome 17[6].

Scientists have detected about 100 imprinted genes in mice, and about half this number in humans. It's not clear if there are genuinely fewer imprinted genes in humans than in mice, or if it's just more difficult to detect them experimentally. Imprinting evolved about 150 million years ago[7], and it really only occurs to a great extent in placental mammals. It isn't found in those classes that can reproduce parthenogenetically.

Imprinting is a complicated system, and like all complex machinery, it can break down. We now know that there are disorders in humans that are caused by problems with the imprinting mechanism.

When imprinting goes bad

Prader-Willi syndrome (PWS) is named after two of the authors of the first description of the condition[8]. PWS affects about one in 20,000 live births. The babies have a low birth weight and their muscles are really floppy. In early infancy, it can be difficult to feed these babies and initially they fail to thrive. This is dramatically reversed by early childhood. The children are constantly hungry, so over-eat to an incredible degree and can become dangerously obese. Along with other characteristic features such as small feet and hands, delayed language development and infertility, the individuals with PWS are often mildly or moderately mentally retarded. They may also have behavioural disturbances, including inappropriate temper outbursts[9].

There's another disorder in humans that affects about the same number of people as PWS. This is called Angelman syndrome (AS), and like PWS it is named after the person who first described the condition[10]. Children with AS suffer from severe mental retardation, small brain size and very little speech. Patients with AS will often laugh spontaneously for no obvious reason, which led to the spectacularly insensitive clinical description of these children as 'happy puppets'[11].

In both PWS and AS, the parents of the affected children are normally perfectly healthy. Research suggested that the basic problem in each of these conditions was likely to be caused by an underlying defect in the chromosomes. Because the parents were unaffected, the defect probably arose during the production of the eggs or the sperm.

In the 1980s, researchers working on PWS used a variety of standard techniques to find the underlying cause of this condition. They looked for regions of the genome that were different between healthy children and those with the disorder. Scientists interested in AS were doing something similar. By the mid-1980s it was becoming clear that both groups were looking at the same part of the genome, a specific stretch on chromosome 15. In both PWS and AS, patients had lost a small, identical section of this chromosome.

But these two disorders are very unlike each other in their clinical presentation. Nobody would ever confuse a patient with PWS with one who was suffering from Angelman's syndrome. How could the same genetic problem – the loss of a key region of chromosome 15 – result in such different symptoms?

In 1989 a group from The Children's Hospital, Boston, showed that the important feature was not just the deletion, but how the deletion was inherited. It's summarised in Figure 8.2. When the abnormal chromosome was inherited from the father, the child had PWS. When the same chromosome abnormality was inherited from the mother, the child had AS[12].

This is a clear case of epigenetic inheritance of a disorder. Children with PWS and AS had exactly the same problem genetically – they were missing a specific region of chromosome 15. The only difference was how they inherited the abnormal chromosome. This is another example of a parent-of-origin effect.

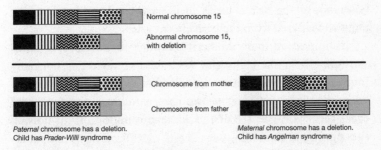

Figure 8.2 Two children may each have the same deletion on chromosome 15, shown schematically by the absence of the horizontally striped box. The phenotype of the two children will be different, depending on how they inherited the abnormal chromosome. If the abnormal chromosome was inherited from their father, the child will develop Prader-Willi syndrome. If the abnormal chromosome was inherited from their mother, the child will develop Angelman syndrome, which is a very different disorder from Prader-Willi.

There's another way in which patients can inherit PWS or AS. Some patients with these disorders have two totally normal copies of chromosome 15. There are no deletions, and no other mutations of any type, and yet the children develop the conditions. To understand how this can be, it's helpful to think back to the mice who inherited both copies of chromosome 11 from one parent. Some of the same researchers who unravelled the story of the PWS deletion showed that in certain examples of this condition, the children have two normal copies of chromosome 15. The trouble is, they've inherited both from their mother, and none from their father. This is known as uniparental disomy – one parent contributing two chromosomes[13]. In 1991, a team from the Institute of Child Health in London showed that some cases of AS were caused by the opposite form of uniparental disomy to PWS. The children had two normal copies of chromosome 15, but had inherited both from their father[14].

This reinforced the notion that PWS and AS are each examples of epigenetic diseases. The children with uniparental disomy of chromosome 15 had inherited exactly the right amount of DNA, they just hadn't inherited it from each parent. Their cells contained all the correct genes, in all the correct amounts, and yet still they suffered from these severe disorders.

It's important that we inherit this fairly small region of chromosome 15 in the right way because this region is normally imprinted. There are genes in this region that are only expressed from either the maternal or the paternal chromosome. One of these genes is called *UBE3A*. This gene is important for normal functioning in the brain, but it's only expressed from the maternally inherited gene in this tissue. But what if a child doesn't inherit a copy of *UBE3A* from its mother? This could happen if both copies of *UBE3A* came from the father, because of uniparental disomy of chromosome 15. Alternatively, the child might inherit a copy of chromosome 15 from its mother which lacked the *UBE3A* gene, because part of the chromosome had been lost.

In these cases, the child can't express UBE3A protein in its brain, and this leads to the development of the symptoms of Angelman syndrome.

Conversely, there are genes that are normally only expressed from the paternal version of this stretch of chromosome 15. This includes a gene called *SNORD116*, but others may also be important. The same scenario applies as for *UBE3A*, but replace the word maternal with paternal. If a child doesn't inherit this region of chromosome 15 from its father, it develops Prader-Willi syndrome.

There are other examples of imprinting disorders in humans. The most famous is called Beckwith-Wiedemann syndrome, again named after the people who first described it in the medical literature[15,16]. This disorder is characterised by over-growth of tissues, so that the babies are born with over-developed muscles including the tongue, and a range of other symptoms[17]. This condition has a slightly different mechanism to the ones described above. When imprinting goes wrong in Beckwith-Wiedemann syndrome, both the maternal and paternal copies of a gene on chromosome 11 get switched on, when normally only the paternally-derived version should be expressed. The key gene seems to be *IGF2*, which codes for the growth factor protein that we met earlier, on mouse chromosome 7. By expressing two copies of this gene, rather than just one, twice as much IGF2 protein as normal is produced and the foetus grows too much.

The opposite phenotype to Beckwith-Wiedemann syndrome is a condition called Silver-Russell syndrome[18,19]. Children with this disorder are characterised by retarded growth before and after birth and other symptoms associated with late development[20]. Most cases of this condition are also caused by problems in the same region of chromosome 11 as in Beckwith-Wiedemann syndrome, but in Silver-Russell syndrome IGF2 protein expression is depressed, and the growth of the foetus is dampened down.

The epigenetic imprint

So, imprinting refers to a situation where there is expression of only one member of a pair of genes, and the expression may be either maternal or paternal. What controls which gene is switched on? It probably isn't surprising to learn that DNA methylation plays a really big role in this. DNA methylation switches genes off. Therefore, if a paternally-inherited region of a chromosome is methylated, the paternally-derived genes in this region will be repressed.

Let's take the example of the *UBE3A* gene which we encountered in the discussion of Prader-Willi and Angelman syndromes. Normally, the copy inherited from the father contains methylated DNA and the gene is switched off. The copy inherited from the mother doesn't have this methylation mark, and the gene is switched on. Something similar happens with *Igf2r* in mice. The paternal version of this is usually methylated, and the gene is inactive. The maternal version is non-methylated and the gene is expressed.

While a role for DNA methylation may not have come as a shock, it may be surprising to learn that it is often not the gene body that is methylated. The part of the gene that codes for protein is epigenetically broadly the same when we compare the maternal and paternal copies of the chromosome. It's the region of the chromosome that *controls* the expression of the gene that is differently methylated between the two genomes.

Imagine a night-time summer party in a friend's garden, beautifully lit by candles scattered between the plants. Unfortunately, this lovely ambience is constantly ruined because the movement of the guests keeps triggering a motion detector on a security system and turning on a floodlight. The floodlight is too high on the wall to be able to cover it, but finally it dawns on the guests that they don't need to cover the light. They need to cover the sensor that is triggering the light's activity. This is very much what happens in imprinting.

The methylation, or lack of it, is on regions known as imprinting control regions (ICRs). In some cases, imprinting control is very straightforward to understand. The promoter region of a gene is methylated on the gene inherited from one parent, and not on the one from the other. This methylation keeps a gene switched off. This works when there is a single gene in a chromosome region that is imprinted. But many imprinted genes are arranged in clusters, all very close to one another in a single stretch on one chromosome. Some of the genes in the cluster will be expressed from the maternally-derived chromosome, others from the paternally-derived one. DNA methylation is still the key feature, but other factors help it to carry out its function.

The imprinting control region may operate over long distances, and certain stretches may bind large proteins. These proteins act like roadblocks in a city, insulating different stretches on a chromosome from one another. This gives the imprinting process an additional level of sophistication, by inserting diversions between different genes. Because of this, an imprinting control region may operate over many thousands of base-pairs, but it doesn't mean that every single gene in those thousands of base-pairs is affected the same way. Different genes in a particular imprinted stretch of chromatin may loop out from their chromosome to form physical associations with each other, so that repressed genes huddle together in a sort of chromatin knot. Activated genes from the same stretch of chromosome may cling together in a different bundle[21].

The impact of imprinting varies from tissue to tissue. The placenta is particularly rich in expression of imprinted genes. This is what we would expect from our model of imprinting as a means of balancing out the demand on maternal resources. The brain also appears to be very susceptible to imprinting effects. It's not so clear why this should be the case. It's harder to reconcile parent-of-origin control of gene expression in the brain with the battle for nutrients we've been considering so far. Professor Gudrun

Moore of University College London has made an intriguing suggestion. She has proposed that the high levels of imprinting in the brain represent a post-natal continuation of the war of the sexes. She has speculated that some brain imprints are an attempt by the paternal genome to promote behaviour in young offspring that will stimulate the mother to continue to drain her own resources, for example by prolonged breast-feeding[22].

The number of imprinted genes is quite low, rather less than 1 per cent of all protein-coding genes. Even this small percentage won't be imprinted in all tissues. In many cells the expression from the maternally and paternally-derived copies will be the same. This is not because the methylation pattern is different between the tissues but because cells vary in the ways that they 'read' this methylation.

The DNA methylation patterns on the imprinting control regions are present in all the cells of the body, and show which parent transmitted which copy of a chromosome. This tells us something very revealing about imprinted regions. They must evade the reprogramming that takes place after the sperm and egg fuse to form the zygote. Otherwise, the methylation modifications would be stripped off and there would be no way for the cell to work out which parent had donated which chromosome. Just as the IAP retrotransposons stay methylated during zygotic reprogramming, mechanisms have evolved to protect imprinted regions from this broad-brush removal of methylation. It's not really very clear how this happens, but it's essential for normal development and health.

You put your imprint on, you take your imprint off …

Yet this presents us with a bit of a problem. If imprinted DNA methylation marks are so stable, how do they change as they are transmitted from parent to offspring? We know that they do, because of Azim Surani's experiments with mice that we encountered in the previous chapter. These showed how methylation of a sequence monitored for experimental purposes changed as it was passed down the generations. This was the experiment that was described using the mice with 'black' and 'white' DNA in the previous chapter.

In fact, once scientists recognised that parent-of-origin effects exist, they predicted that there must be a way to reset the epigenetic marks, even before they knew what these marks were. Let's consider chromosome 15, for example. I inherited one copy from my mother and one from my father. The *UBE3A* imprinting control region from my mother was unmethylated, whereas the same region on the chromosome from my father was methylated. This ensured appropriate expression patterns of UBE3A protein in my brain.

When my ovaries produce eggs, each egg inherits just one copy of chromosome 15, which I will pass on to a child. Because I'm a woman, each copy of chromosome 15 must carry a maternal mark on *UBE3A*. But one of my copies of chromosome 15 has been carrying the paternally-derived mark I inherited from my father. The only way I can make sure that I pass on chromosome 15 with the correct maternal mark to my children is if my cells have a way of removing the paternal mark and replacing it with a maternal one.

A very similar process would have to take place when males produce sperm. All maternally-derived modifications would need to be stripped off the imprinted genes, and paternally derived ones put on in their place. This is indeed exactly what happens. It's a very restricted process which only takes place in the cells that give rise to the germ line.

The general principle is shown diagrammatically in Figure 8.3.

Following fusion of the egg and sperm the blastocyst forms, and most regions of the genome become reprogrammed. The cells begin to differentiate, forming the precursors to the placenta

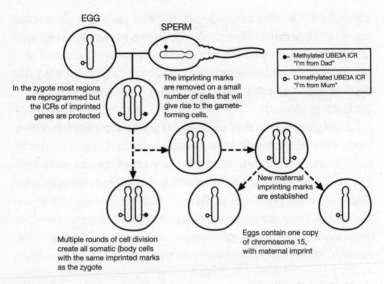

Figure 8.3 Diagram showing how the somatic cells arising from a fertilised zygote all carry the same DNA methylation patterns as each other at imprinted genes, but the imprinting methylation is removed and then re-established in the germ cells. This ensures that females only pass on maternal marks to their offspring, and males only pass on paternal ones.

and also the various cell types of the body. So, at this point the cells that had been part of the ICM are all marching to the developmental drumbeat, heading down the various troughs in Waddington's epigenetic landscape. But a very small number (less than 100) begin to march to a different beat. In these cells a gene called *Blimp1* switches on. Blimp1 protein sets up a new cascade in signalling, which stops the cells heading towards their somatic dead-ends. These cells start travelling back up Waddington's trenches[23]. They also lose the imprinted marks which told the cell which parent donated which of a pair of chromosomes.

The tiny population of cells that carry out this process are know as the primordial germ cells. It's these cells that will ulti-mately settle in the developing gonads (testicles or ovaries) and act as the stem cells that produce all the gametes (sperm or eggs respectively). In the stage described in the previous paragraph, the primordial germ cells are reverting to a state more like that of the cells of the inner cell mass (ICM). Essentially, they are becoming pluripotent, and potentially able to code for most of the tissue types in the body. This phase is fleeting. The primordial germ cells quickly get diverted into a new developmental pathway where they differentiate to form stem cells that will give rise to eggs or sperm. To do so, they gain a new set of epigenetic modifications. Some of these modifications are ones that define cellular identity, i.e. switch on the genes that make an egg an egg. But a small number are the ones that serve as parent-of-origin marks, so that in the next generation the imprinted regions of the genome can be recognised with respect to their parent-of-origin.

This seems horribly complicated. If we follow the path from the sperm that fertilised the egg to a new sperm being formed in male offspring, the sequence goes like this:

1. The sperm that enters the egg has epigenetic modifications on it;
2. The epigenetic modifications get taken off, except at the imprinted regions (in the immediate post-fertilisation zygote);
3. Epigenetic modifications get put on (as the cells of the ICM begin to specialise);
4. The epigenetic modifications get taken off, including at the imprinted regions (as the primordial germ cells break away from the somatic differentiation pathway);
5. Epigenetic modifications get put on (as the sperm develops).

This could seem like an unnecessarily complicated way to get back to where we started from, but it's essential.

The modifications that make a sperm a sperm, or an egg an egg, have to come off at stage 2 or the zygote wouldn't be totipotent. Instead it would have a genome that was half-programmed to be an egg and half-programmed to be a sperm. Development wouldn't be possible if the inherited modifications stayed on. But to create primordial germ cells, some of the cells from the differentiating ICM have to lose their epigenetic modifications. This is so they can become temporarily more pluripotent, lose their imprinting marks and transfer across into the germ cell lineage.

Once the primordial germ cells have been diverted, epigenetic modifications again get attached to the genome. This is partly because pluripotent cells are potentially extremely dangerous as a multi-cellular organism develops. It might seem like a great idea to have cells in our body that can divide repeatedly and give rise to lots of other cell types, but it's not. Those sorts of cells are the type that we find in cancer. Evolution has favoured a mechanism where the primordial germ cells can regain pluripotency for a period, but then this pluripotency is re-suppressed by epigenetic modifications. Coupled with this, the wiping out of the imprints means that chromosomes can be marked afresh with their parent-of-origin.

Occasionally this process of setting up the new imprints on the progenitors of egg or sperm can go wrong. There are cases of Angelman syndrome and Prader-Willi syndrome where the imprint has not been properly erased during the primordial germ cell stage[24]. For example, a woman may generate eggs where chromosome 15 still has the paternal mark on it that she inherited from her father, rather than the correct maternal mark. When this egg is fertilised by a sperm, both copies of chromosome 15 will function like paternal chromosomes, and create a phenotype just like uniparental disomy.

Research is ongoing into how all these processes are controlled. We don't fully understand how imprints are protected from reprogramming following fusion of the egg and the sperm, nor how they lose this protection during the primordial germ cell stage. We're also not entirely sure how imprints get put back on in the right place. The picture is still quite foggy, although details are starting to emerge from the haze.

Part of this may involve the small percentage of histones that are present in the sperm genome. Many of these are located at the imprinting control regions, and may protect these regions from reprogramming when the sperm and the egg fuse[25]. Histone modifications also play a role in establishing 'new' imprints during gamete production. It seems to be important that the imprinting control regions lose any histone modifications that are associated with switching genes on. Only then can the permanent DNA methylation be added[26]. It's this permanent DNA methylation that marks a gene with a repressive imprint.

Dolly and her daughters

The reprogramming events in the zygote and in primordial germ cells impact on a surprising number of epigenetic phenomena. When somatic cells are reprogrammed in the laboratory using the Yamanaka factors, only a tiny percentage of them form iPS cells. Hardly any seem to be exactly the same as ES cells, the genuinely pluripotent cells from the inner cell mass of the blastocyst. A group in Boston, based at Massachusetts General Hospital and Harvard University, assessed genetically identical iPS and ES cells from mice. They looked for genes that varied in expression between the two types of cells. The only major differences in expression were in a chromosomal region known as *Dlk1-Dio3*[27]. A few iPS cells expressed the genes in this region in a way that was very similar to how the ES cell did this. These were the best iPS cells for forming all the different tissues of the body.

Dlk1-Dio3 is an imprinted region on chromosome 12 of the mouse. It's perhaps not surprising that an imprinted region turned out to be so important. The Yamanaka technique triggers the reprogramming process that normally occurs when a sperm fuses with an egg. Imprinted regions of the genome are resistant to reprogramming in normal development. It is likely that they present too high a barrier to reprogramming in the very artificial environment of the Yamanaka method.

The *Dlk1-Dio3* region has been of interest to researchers for quite some time. In humans, uniparental disomy in this region is associated with growth and developmental defects, amongst other symptoms[28]. This region has also been shown to be critical for the prevention of parthenogenesis, at least in mice. Researchers from Japan and South Korea genetically manipulated just this region of the genome in mice. They reconstructed a fertilised egg with two female pronuclei. The *Dlk1-Dio3* region in one of the pronuclei had been altered so that it carried the equivalent of a paternal rather than maternal imprint. The live mice that were born were the first example of a placental mammal with two maternal genomes[29].

The reprogramming that occurs in the primordial germ cells isn't completely comprehensive. It leaves the methylation on some IAP retrotransposons more or less intact. The DNA methylation level of the $Axin^{Fu}$ retrotransposon in sperm is the same as it is in the body cells of this strain of mice. This shows that the DNA methylation was not removed when the PGCs were reprogrammed, even though most other areas of the genome did lose this modification. This resistance of the $Axin^{Fu}$ retrotransposon to both rounds of epigenetic reprogramming (in the zygote and in the primordial germ cells) provides a mechanism for the transgenerational inheritance of the kinked tail trait that we met in earlier chapters.

We know that not all transgenerational inheritance happens in the same way. In the *agouti* mouse the phenotype is transmitted via the mother, but not via the father. In this case, the DNA methylation on the IAP retrotransposon is removed in both males and females during normal primordial germ cell reprogramming. However, mothers whose retrotransposon originally carried DNA methylation pass on a specific histone mark to their offspring.

This is a repressive histone modification and it acts as a signal to the DNA methylation machinery. This signal attracts the enzymes that put the repressive DNA methylation onto a specific region on a chromosome. The final outcome is the same – the DNA methylation in the mother is restored in the offspring. Male *agouti* mice don't pass on either DNA methylation or repressive histone modifications on their retrotransposon, which is why transmission of the phenotype only occurs through the maternal line[30].

This is a slightly more indirect method of transmitting epigenetic information. Instead of direct carry-over of DNA methylation, an intermediate surrogate (a repressive histone modification) is used instead. This is probably why the maternal transmission of the *agouti* phenotype is a bit 'fuzzy'. Not all offspring are exactly the same as the mother, because there is a bit of 'wriggle-room' in how DNA methylation gets re-established in the offspring.

In the summer of 2010, there were reports in the British press about cloned farm animals. Meat that had come from the offspring of a cloned cow had entered the human food chain[31]. Not the cloned cow itself, just its offspring, created by conventional animal breeding. Although there were a few alarmist stories about people unwittingly eating 'Frankenfoods', the coverage in the mainstream media was pretty balanced.

To some extent, this was probably because of a quite intriguing phenomenon, which has allayed certain fears originally held by scientists about the consequences of cloning. When cloned animals breed, the offspring tend to be healthier than the original clones. This is almost certainly because of primordial germ cell reprogramming. The initial clone was formed by transfer of a somatic nucleus into an egg. This nucleus only went through the first round of reprogramming, the one that normally happens when a sperm fertilises an egg. The likelihood is that this epigenetic reprogramming wasn't entirely effective – it's a big ask to get an egg to reprogram a 'wrong' nucleus. This is likely to be the reason why clones tend to be unhealthy.

When the cloned animals breed, they pass on either an egg or a sperm. Before the clone produced these gametes, its primordial cells underwent the second round of reprogramming, as part of the normal primordial germ cell pathway. This second reprogramming stage seems to reset the epigenome properly. The gametes lose the abnormal epigenetic modifications of their cloned parent. Epigenetics explains why cloned animals have health issues, but also explains why their offspring don't. In fact, the offspring are essentially indistinguishable from animals produced naturally.

Assisted reproductive technologies in humans (such as in vitro fertilisation) share certain technical aspects with some of the methods used in cloning. In particular, pluripotent nuclei may be transferred between cells, and cells are cultured in the laboratory before being implanted in the uterus. There is a substantial amount of controversy in the scientific journals about the abnormality rates from these procedures[32]. Some authors claim there is an increased rate of imprinting disorders in pregnancies from assisted reproductive technologies. This would imply that procedures such as culturing fertilised eggs outside the body may disrupt the delicately poised pathways that control reprogramming, especially of imprinted regions. It's important to note, however, that there is no consensus yet on whether this really is a clinically relevant issue.

All the reprogramming of the genome in early development has multiple effects. It allows two highly differentiated cell types to fuse and form one pluripotent cell. It balances out the competing demands of the maternal and paternal genomes, and ensures that this balancing act can be re-established in every generation. Reprogramming also prevents inappropriate epigenetic modifications being passed from parent to offspring. This means that even if cells have accumulated potentially dangerous epigenetic changes, these will be removed before they are passed on.

This is why we don't normally inherit acquired characteristics. But there are certain regions of the genome, such as IAP retrotransposons, that are relatively resistant to reprogramming. If we want to work out how certain acquired characteristics – responses to vinclozolin or responses to paternal nutrition, for example – get transmitted from parent to offspring, these IAP retrotransposons might be a good place to start looking.

Chapter 9

Generation X

The sound of a kiss is not so loud as that of a cannon, but its echo lasts a great deal longer.
Oliver Wendell Holmes

At a purely biological, and especially an anatomical level, men and women are different. There are ongoing debates about whether or not certain behaviours, ranging from aggression to spatial processing, have a biological gender bias. But there are certain physical characteristics that are linked unequivocally to gender. One of the most fundamental differences is in the reproductive organs. Women have ovaries, men have testicles. Women have a vagina and a uterus, men have a penis.

There is a clear biological basis to this, and perhaps unsurprisingly, it's all down to genes and chromosomes. Humans have 23 pairs of chromosomes in their cells, and inherited one of each pair from each parent. Twenty-two of these pairs (imaginatively named chromosomes 1 to 22) are called autosomes and each member of a specific pair of autosomes looks very similar. By 'looks' we mean exactly that. At a certain stage in cell division the DNA in chromosomes becomes exceptionally tightly coiled up. If we use the right techniques we can actually see chromosomes down a microscope. These chromosomes can be photographed. In pre-digital days, clinical geneticists literally used to cut out the pictures of the individual chromosomes with a pair of scissors and rearrange them in pairs to create a nice orderly picture. These days the image processing can be carried out by a computer, but in either case the result is a picture of all the chromosomes in a cell. This picture is called a karyotype.

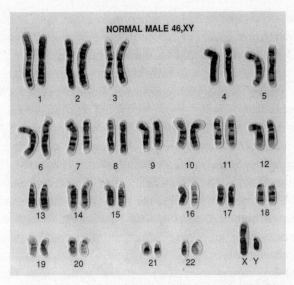

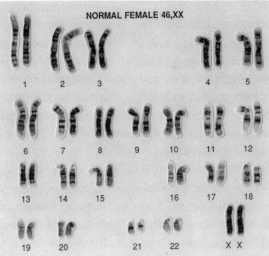

Figure 9.1 Karyotype of all the chromosomes in a male (top) and female (bottom) somatic cell. Note that the female cell contains two X chromosomes and no Y chromosome; the male cell contains one X chromosome and one Y chromosome. Note also the substantial difference in size between the X and Y chromosomes. Photos: Wessex Reg Genetics Centre/Wellcome Images.

Karyotype analysis is how scientists originally discovered that there were three copies of chromosome 21 in the cells of people with Down's syndrome. This is known as trisomy 21.

When we produce a human karyotype from a female, there are 23 pairs of identical chromosomes. But if we create a human karyotype from a male, the picture is different, as we can see in Figure 9.1. There are 22 obvious pairs – the autosomes – but there are two chromosomes left over that don't look like each other at all. One is very large, one exceptionally small. These are called the sex chromosomes. The large one is called X, and the small one is called Y. The notation to describe the normal chromosome constitution of human males is 46, XY. Females are described as 46, XX because they don't have a Y chromosome, and instead have two X chromosomes.

The Y chromosome carries very few active genes. There are only between 40 and 50 protein-coding genes on the Y chromosome, of which about half are completely male-specific. The male-specific genes only occur on the Y chromosome, so females have no copies of these. Many of these genes are required for male-specific aspects of reproduction. The most important one in terms of sex determination is a gene called *SRY*. SRY proteins activate a testis-determining pathway in the embryo. This leads to production of testosterone, the archetypal 'male' hormone, which then masculinises the embryo.

Occasionally, individuals who phenotypically appear to be girls have the male 46, XY karyotype. In these cases the *SRY* gene is often inactive or deleted and consequently the foetus develops down the default female pathway[1]. Sometimes, the other scenario arises. Individuals who phenotypically appear to be boys can have the typically female karyotype of 46, XX. In these cases a tiny section of the Y chromosome containing the *SRY* gene has often transferred onto another chromosome during formation of sperm in the father. This is enough to drive masculinisation of the foetus[2]. The region of the Y chromosome that was transferred was too small to be detected by the karyotyping process.

The X chromosome is very different. The X chromosome is extremely large and carries about 1300 genes. A disproportionate number of these genes are involved in brain function. Many are also required for various stages in formation of the ovaries or the testes, and for other aspects of fertility in both males and females[3].

Getting the dose right

So, about 1300 genes on the X chromosome. That creates an interesting problem. Females have two X chromosomes but males only have one. That means that for these 1300 genes on the X, females have two copies of each gene but males only have one. We might speculate from this that female cells would produce twice the amounts of proteins from these genes (referred to as X-linked genes) as males.

But our knowledge of disorders like Down's syndrome makes this seem rather unlikely. Having three copies of chromosome 21 (instead of the normal two) results in Down's syndrome, which is a major disorder in those individuals who are born with the condition. Trisomies of most other chromosomes are so severe that children are never born with these conditions, because the embryos cannot develop properly. For example, no child has ever been born who has three copies of chromosome 1 in all their cells. If the 50 per cent increase in gene expression from an autosome can cause such problems in trisomic conditions, how do we explain the X chromosome scenario? How is it possible for females to survive when they have twice as many X chromosome genes as males? Or, to put it the other way – why are males viable if they only have half as many X chromosome genes as females?

The answer is that expression of X-linked genes is actually pretty much the same in males and females, despite the different number of chromosomes, a phenomenon called dosage compensation. The XY system of sex determination doesn't exist in other animal classes so X chromosome dosage compensation is limited to placental mammals.

In the early 1960s a British geneticist called Mary Lyon postulated how dosage compensation would occur at the X chromosome. These were her predictions:

1. Cells from the normal female would contain only one active X chromosome;
2. X inactivation would occur early in development;
3. The inactive X could be either maternally or paternally derived, and the inactivation would be random in any one cell;
4. X inactivation would be irreversible in a somatic cell and all its descendants.

These predictions have proven remarkably prescient[4,5]. So prescient, in fact, that many textbooks refer to X inactivation as Lyonisation. We'll take the predictions one at a time:

1. Individual cells from a normal female do indeed only express genes from one X chromosome copy – the other copy is, effectively, shut down;
2. X inactivation occurs early in development, at the stage when the pluripotent cells of the embryonic inner cell mass are beginning to differentiate into different lineages (near the top of Waddington's epigenetic landscape);
3. On average, in 50 per cent of cells in a female the maternally derived X chromosome is shut down. In the other 50 per cent of cells it's the chromosome inherited from Dad which gets inactivated;

4. Once a cell has switched off one of a pair of X chromosomes, that particular copy of the X stays switched off in all the daughter cells for the rest of that woman's life, even if she lives to over 100 years of age.

The X chromosome isn't inactivated by mutation; it keeps its DNA sequence entirely intact. X inactivation is the epigenetic phenomenon par excellence.

X inactivation has proven to be a remarkably fertile research field. Some of the mechanisms involved have turned out to have parallels in a number of other epigenetic and cellular processes. The consequences of X inactivation have important implications for a number of human disorders and for therapeutic cloning. Yet even now, 50 years on from Mary Lyon's ground-breaking work, there remain a number of mysteries about how X inactivation actually takes place.

The more we ponder X inactivation, the more extraordinary it appears. For a start, the inactivation is only on the X chromosome, not on any of the autosomes, so the cell must have a way of distinguishing X chromosomes and autosomes from one another. Furthermore, the inactivation in the X doesn't just affect one or a few genes, such as occurs in imprinting. No, in X inactivation, over 1,000 genes are turned off, for decades.

Think of a car manufacturer, with a factory in Japan and another in Germany. Imprinting is the equivalent of a few changes in specification for the different markets. The German factory may switch on the machine that installs the heater on the steering wheel and switch off the robot that inserts the automatic air freshener, whilst the Japanese factory does the opposite. X inactivation is the equivalent of shutting down and mothballing one factory, never to be re-opened unless the company is bought by a new manufacturer.

Random inactivation

The other major difference between X-inactivation and imprinting is that there is no parent-of-origin effect in X imprinting. In somatic cells, it doesn't matter if an X chromosome was inherited from your mother or your father. Each has a 50 per cent chance of being inactivated. The reason why this is the case makes complete evolutionary sense.

Imprinting is about balancing out the competing demands of the maternal and paternal genomes, especially during development. The imprinting mechanisms that have evolved are specifically targeted at individual genes, or small clusters of genes, that particularly influence foetal growth. There are, after all, only 50–100 imprinted genes in the mammalian genome.

But X inactivation operates on a much greater scale. It's a mechanism for switching off over 1,000 genes, en masse and permanently. A thousand genes is a lot, about 5 per cent of the total number of protein-coding genes, so there's always a possibility that any given gene on an X chromosome may have a mutation. Figure 9.2 compares the outcomes of imprinted X inactivation on the left, with random X inactivation on the right. For clarity, the diagram just exemplifies a mutation in a paternally inherited gene, with imprinted inactivation of the maternally derived X chromosome.

By using *random* X inactivation, cells are able to minimise the effects of mutations in X-linked genes.

It's important to bear in mind that the inactive X really is inactive. Almost all the genes are permanently shut off and this inactivation cannot normally be broken. When we refer to the active X chromosome, we are using slightly ambiguous shorthand. It doesn't mean that every gene on that X is active all the time in every cell. Rather, the genes have the potential to be active. They are subject to all the normal epigenetic modifications and controls on gene expression, so that selected genes are switched on or

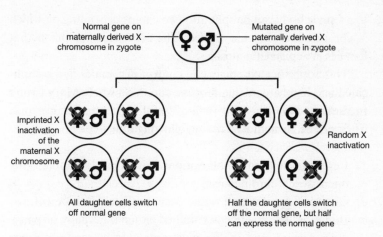

Figure 9.2   Each circle represents a female cell, containing two X chromosomes. The X chromosome inherited from the mother is indicated by the female symbol. The X chromosome inherited from the father is indicated by the male symbol, and contains a mutation, denoted by the white square notch. The left hand side of the diagram demonstrates that imprinted inactivation of the maternally derived X chromosome would result in all cells of the body expressing only the X chromosome carrying the mutation, which was inherited from the father. On the right hand side, the X chromosomes are randomly inactivated, independent of their parent-of-origin. As a result, on average, half of the somatic cells will express the normal version of the X chromosome. This makes random X inactivation a less risky evolutionary strategy than imprinted X inactivation.

off in a controlled manner, in response to developmental cues or environmental signals.

Women really are more complicated than men

One interesting consequence of X inactivation is that (epigenetically) females are more complicated than males. Males only have one X chromosome in their cells, so they don't carry out X inactivation. But females randomly inactivate an X chromosome in all their cells. Consequently, at a very fundamental level, all cells in a female body can be split into two camps depending on which X chromosome they inactivated. The expression for this is that females are epigenetic mosaics.

This sophisticated epigenetic control in females is a complicated and highly regulated process, and that's where Mary Lyon's predictions have provided such a useful conceptual framework. They can be paraphrased as the following four steps:

1. Counting: cells from the normal female would contain only one active X chromosome;
2. Choice: X inactivation would occur early in development;
3. Initiation: the inactive X could be either maternally or paternally derived, and the inactivation would be random in any one cell;
4. Maintenance: X inactivation would be irreversible in a somatic cell and all its descendants.

Unravelling the mechanisms behind these four processes has kept researchers busy for nearly 50 years, and this effort is continuing today. The processes are incredibly complicated and sometimes involve mechanisms that had barely been imagined by any scientists. That's not really surprising, because Lyonisation is quite extraordinary – X inactivation is a procedure where a cell treats two identical chromosomes in diametrically opposite and mutually exclusive ways.

Experimentally, X inactivation is challenging to investigate. It is a finely balanced system in cells, and slight variations in technique may have a major impact on the outcome of experiments. There's also considerable debate about the most appropriate species to study. Mouse cells have traditionally been used as the experimental system of choice, but we are now realising that mouse and human cells aren't identical with respect to X inactivation[6]. However, even allowing for these ambiguities, a fascinating picture is beginning to emerge.

Counting chromosomes

Mammalian cells must have a mechanism to count how many X chromosomes they contain. This prevents the X chromosome from being switched off in male cells. The importance of this was shown in the 1980s by Davor Solter. He created embryos by transferring male pronuclei into fertilised eggs. Males have an XY karyotype, and when they produce gametes each individual sperm will contain either an X or a Y. By taking pronuclei from different sperm and injecting them into 'empty' eggs, it was possible to create XX, XY or YY zygotes. None of these resulted in live births, because a zygote requires both maternal and paternal inputs, as we have already seen. But the results still told us something very interesting, and are summarised in Figure 9.3.

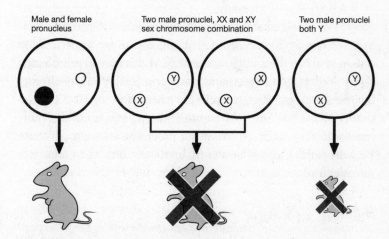

Figure 9.3   Donor egg reconstitution experiments were performed in which the donor egg received a male and female pronucleus or two pronuclei from males. Just as in Figure 7.2, the embryos derived from two male pronuclei failed to develop to term. When the nuclei each contained a Y chromosome, and no X chromosome, the embryos failed at a very early stage. Embryos derived from two male pronuclei where at least one contained an X chromosome developed further before they also died.

The earliest loss of embryos occurred in those that had been reconstituted from two male pronuclei which each contained a Y chromosome as the sole sex chromosome[7]. In these embryos there was no X chromosome at all, and this was associated with exceptionally early developmental failure. This shows that the X chromosome is clearly essential for viability. This is why male (XY) cells need to be able to count, so that they can recognise that they only contain one X, and thus avoid inactivating it. Turning off the solitary X would be disastrous for the cell.

Having counted the number of X chromosomes, there must be a mechanism in female cells by which one X is randomly selected for inactivation. Having selected a chromosome, the cell starts the inactivation procedure.

X inactivation happens early in female embryogenesis, as the cells of the ICM begin to differentiate into the different cell types of the body. Experimentally, it is difficult to work on the small number of cells available from each blastocyst so researchers typically use female ES cells. Both X chromosomes are active in these cells, just like in the undifferentiated ICM. It's easy to roll ES cells down Waddington's epigenetic landscape, just by subtly altering the conditions in which the cells are cultured in the lab. Once we change the conditions to encourage the female ES cells to differentiate, they begin to inactivate an X chromosome. Because ES cells can be grown in almost limitless numbers in labs, this provides a convenient model system for studying X inactivation.

Painting an X-rated picture

Initial insights into X inactivation came from studying mice and cell lines with structurally rearranged chromosomes. In some of these studies, various sections of an X chromosome were missing. Depending on which parts were missing, the X chromosome did or did not inactivate normally. In other studies, sections had come off the X chromosome and attached themselves onto an autosome. Depending on which part of the X chromosome had transferred, this could result in switching off the structurally abnormal autosome[8,9].

These experiments showed that there was a region on the X chromosome that was vitally important for X inactivation. This region was dubbed the X Inactivation Centre. In 1991 a group from Hunt Willard's lab at Stanford University in California showed that the X Inactivation Centre contained a gene that they called *Xist*, after X-inactive (X_i) specific transcript[10]. This gene was only expressed from the *inactive* X chromosome, not from the active one. Because the gene was only expressed from one of the two X chromosomes, this made it an attractive candidate as the controller of X inactivation, where two identical chromosomes behave non-identically.

Attempts were made to identify the protein encoded by the *Xist* gene[11] but by 1992 it was clear that there was something very strange going on. The *Xist* gene was transcribed to form RNA copies. The RNA was processed just like any other RNA. It was spliced, and various structures were added to each end of the transcript to improve its stability. So far, so normal. But before RNA molecules can code for protein, they have to move out of the nucleus and into the cytoplasm of the cell. This is because the ribosomes – the intracellular factories that join amino acids into long protein chains – are only found in the cytoplasm. But the *Xist* RNA never moved out of the nucleus, which meant it could never generate a protein[12,13].

This at least cleared up one thing that had puzzled the scientific community when the *Xist* gene was first identified. Mature *Xist* RNA is a long molecule, of about 17,000 base-pairs (17kb). One amino acid is coded for by a three base-pair codon, as described in Chapter 3. Therefore, in theory, the 17,000 base-pairs of *Xist* should be able to code for a protein of about 5,700 amino acids. But when researchers analysed the *Xist* sequence with protein prediction programs, they simply couldn't see how it could encode anything this long. There were stop codons (which signal the end of a protein) all through the *Xist* sequence and the longest predicted run without stop codons was only enough to code for 298 amino acids (894 base-pairs[14]). Why would a gene have evolved which created a 17kb transcript, but only used about 5 per cent of this to encode protein? That would be a very inefficient use of energy and resources in a cell.

But since *Xist* never actually leaves the nucleus, its lack of potential protein coding is irrelevant. *Xist* doesn't act as a messenger RNA (mRNA) that transmits the code for a protein. It is a class of molecule called a non-coding RNA (ncRNA). *Xist* may not code for protein, but this doesn't mean it has no activity. Instead, the *Xist* ncRNA itself acts as a functional molecule, and it is critical for X inactivation.

Back in 1992 ncRNAs were a real novelty, and only one other was known at the time. Even now, there is something very unusual about *Xist*. It's not just that it doesn't leave the nucleus. *Xist* doesn't even leave the chromosome that produces it. When ES cells begin to differentiate, only one of the chromosomes produces *Xist* RNA. This is the chromosome that will be the inactive one. *Xist* doesn't move away from the chromosome that produced it. Instead, it binds to the chromosome and starts to spread out along it.

Xist is often described as 'painting' the inactive X and it's a very good description. Let's revert yet again to our analogy of the DNA code as a script. This time we'll imagine that the script is written on a wall, maybe it's an inspiring poem or speech in a classroom. At the end of the summer term the school building closes down and is sold for conversion to apartments. The decorators arrive and paint over the script. Now there's nothing to tell the new residents to 'play up and play the game', or exactly how they should 'meet with Triumph and Disaster'. But the instructions are actually still there, they're just hidden from view.

When *Xist* binds over the X chromosome that produced it, it induces a kind of creeping epigenetic paralysis. It covers more and more genes, switching them off. It first seems to do this by acting as a barrier between the genes and the enzymes that normally copy them into mRNA. But as the X inactivation gets better established, it changes the epigenetic modifications on the chromosome. The histone modifications that normally turn genes on are removed. They are replaced by repressive histone modifications that turn genes off.

Some of the normal histones are removed altogether. Histone H2A is replaced by a related but subtly different molecule called macroH2A, strongly associated with gene repression. The promoters of genes undergo DNA methylation, an even more stringent way of turning the genes off. All these changes lead to binding of more and more repressor molecules, coating the DNA on the inactive X and making it less and less accessible to the enzymes that transcribe genes. Eventually, the DNA on the X chromosome gets incredibly tightly wound up, like a giant wet towel being turned at each end, and the whole chromosome moves to the edge of the nucleus. By this stage most of the X chromosome is completely inactive, except for the *Xist* gene, which is a little pool of activity in the midst of a transcriptional desert[15].

Whenever a cell divides, the modifications to the inactive X are copied over from mother cell to daughter cell, and so the same X remains inactivated in all subsequent generations of that starter cell.

While the effects of *Xist* are amazing, the description above still leaves a lot of questions unanswered. How is *Xist* expression controlled? Why does it switch on when ES cells start to differentiate? Is *Xist* only functional when it's in female cells, or could it act in males cells too?

The power of a kiss

The last question was first addressed in the lab of Rudi Jaenisch, whom we met in the context of iPS cells and Shinya Yamanaka's work in Chapter 2. In 1996, Professor Jaenisch and his colleagues created mice carrying a genetically engineered version of the X Inactivation Centre (an X Inactivation Centre transgene). This was 450kb in size, and included the *Xist* gene plus other sequences on either side. They inserted this into an autosome (non-sex chromosome), created male mice carrying this transgene, and studied ES cells from these mice. The male mice only contained one normal X chromosome, because they have the XY karyotype. However, they had two X Inactivation Centres. One was on the normal X chromosome, and one was on the transgene on the autosome. When the researchers differentiated the ES cells from these mice, they found that *Xist* could be expressed from either of the X Inactivation Centres. When *Xist* was expressed, it inactivated the chromosome from which it was expressed, even if this was the autosome carrying the transgene[16].

These experiments showed that even cells that are normally male (XY) can count their X chromosomes. Actually, to be more specific, it showed they could count their X Inactivation Centres. The data also demonstrated that the critical features for counting, choosing and initiation were all present in the 450kb of the X Inactivation Centre around the *Xist* gene.

We know a bit more now about the mechanism of chromosome counting. Cells don't normally count their autosomes. Both copies of chromosome 1, for example, operate independently. But we know that the two copies of the X chromosome in a female ES cell somehow communicate with each other. When X inactivation is getting going, the two X chromosomes in a cell do something very weird.

They kiss.

That's a very anthropomorphic way of describing the event, but it's a pretty good description. The 'kiss' only lasts a couple of hours or so, and it's startling to think this sets a pattern that can persist in cells for the next hundred years, if a woman lives that long. This chromosomal smooch was first shown in 1996 by Jeannie Lee, who started out as a post-doctoral researcher in Rudi Jaenisch's lab, but who is now a professor in her own right at Harvard Medical School, where she was one of the youngest professors ever appointed. She showed that essentially the two copies of the X find each other and make physical contact. This physical contact is only over a really small fraction of the whole chromosome, but it's essential for triggering inactivation[17]. If it doesn't happen, then the X chromosome assumes it is all alone in the cell, *Xist* never gets switched on, and there is no X inactivation. This is a key stage in chromosome counting.

It was Jeannie Lee's lab that also identified one of the critical genes that controls *Xist* expression[18]. DNA is double-stranded, with the bases in the middle holding the strands together. Although we often envisage it as looking like a railway track, it might be better to think of it as two cable cars, running in opposite directions. If we use this metaphor, then the X Inactivation Centre looks a bit like Figure 9.4.

There is another non-coding RNA, about 40kb in length, in the same stretch of DNA as *Xist*. It overlaps with *Xist* but is on the opposite strand of the DNA molecule. It is transcribed into RNA in the opposite direction to *Xist* and is referred to as an antisense transcript. Its name is *Tsix*. The eagle-eyed reader will

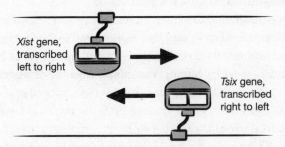

Figure 9.4   The two strands of DNA at a specific location on the X chromosome can each be copied to create mRNA molecules. The two backbones are copied in opposite directions to each other, allowing the same region of the X chromosome to produce *Xist* RNA or *Tsix* RNA.

notice that *Tsix* is *Xist* backwards, which has an unexpectedly elegant logic to it.

This overlap in location between *Tsix* and *Xist* is really significant in terms of how they interact, but it makes it exceedingly tricky to perform conclusive experiments. That's because it's very difficult to mutate one of the genes without mutating its partner on the opposite strand, a sort of collateral damage. Despite this, considerable strides have been made in understanding how *Tsix* influences *Xist*.

If an X chromosome expresses *Tsix*, this prevents *Xist* expression from the same chromosome. Oddly enough, it may be the simple action of transcribing *Tsix* that prevents the *Xist* expression, rather than the *Tsix* ncRNA itself. This is analogous to a mortice lock. If I lock a mortice from the inside of my house and leave the key in the lock, my partner can't unlock the door from the outside of the house. I don't need to keep locking the door, just having the key in there is enough to stop the action of someone on the other side. So, when *Tsix* is switched on, *Xist* is switched off and the X chromosome is active.

This is the situation in ES cells, where both X chromosomes are active. Once the ES cells begin to differentiate, one of the pair stops expressing *Tsix*. This allows expression of *Xist* from that X chromosome, which drives X inactivation.

Tsix alone is probably not enough to keep *Xist* repressed. In ES cells, the proteins Oct4, Sox2 and Nanog bind to the first intron of *Xist* and suppress its expression[19]. Oct4 and Sox2 were two of the four factors used by Shinya Yamanaka when he reprogrammed somatic cells to the pluripotent iPS cell type. Subsequent experiments showed that Nanog (named after the mythical Celtic land of everlasting youth) can also work as a reprogramming factor. Oct4, Sox2 and Nanog are highly expressed in undifferentiated cells like ES cells, but their levels fall as cells start to differentiate. When this happens in differentiating female ES cells, Oct4, Sox2 and Nanog stop binding to the *Xist* intron. This removes some of the barriers to *Xist* expression. Conversely, when female somatic cells are reprogrammed using the Yamanaka approach, the inactive X chromosome is reactivated[20]. The only other time the inactive X is reactivated is during the formation of primordial germ cells in development, which is why the zygote starts out with two active X chromosomes.

We are still a bit vague as to why X inactivation is so mutually exclusive between the pair of chromosomes. One theory is that it's all down to what happens when the X chromosomes kiss. This happens at a developmental point where *Tsix* levels are starting to fall, and the levels of the Yamanaka factors are also declining. The theory is that the pair of chromosomes reaches some sort of compromise. Rather than each ending up with a sub-optimal amount of non-coding RNAs and other factors, the binding molecules all get shunted together onto one of the pair. There's not a great deal of clarity on how this happens. It could be that one of the pair of chromosomes just by chance carries slightly more of a key factor than the other. This makes it slightly more attractive to certain proteins. Complexes may build up in a self-sustaining way, so that the more of a complex one chromosome starts with, the more it can drag off its partner. The rich get richer, the poor get poorer …

It's quite remarkable how many gaps remain in our understanding of X inactivation, 50 years after Mary Lyon's formative work. We don't even really understand how the *Xist* RNA ends up coating the chromosome from which it is expressed, or how it recruits all those negative repressive epigenetic enzymes and modifications. So perhaps it's timely to move off the shifting sands and step back onto more solid ground.

Let's return to this statement from earlier in the chapter: 'Once a cell has switched off one of a pair of X chromosomes, that particular copy of the X stays switched off in all the daughter cells for the rest of that woman's life, even if she lives to over a hundred years of age.' How do we know that? How can we be so certain that X inactivation is stable in somatic cells? It is now possible to perform genetic manipulation to show this in species like mice. But long before that became feasible scientists were already pretty certain this was the case. For this piece of information we thank not mice, but cats.

Learning from the epigenetic cat

Not just any old cats, but specifically tortoiseshell ones. You probably know how to recognise a classic tortoiseshell cat. It's the one that's a mixture of black and ginger splodges, sometimes on a white background. The colour of each hair in a cat's coat is caused by cells called melanocytes that produce pigment. Melanocytes are found in the skin, and develop from special stem cells. When melanocyte stem cells divide, the daughter cells stay close to each other, forming a little patch of clonal cells from the same parent stem cell.

Now, here's an amazing thing: if a cat's colour is tortoiseshell, it's a female.

There is a gene for coat colour that encodes either black pigment or orange pigment. This gene is carried on the X chromosome. A cat may receive the black version of the gene on the X chromosome inherited from her mother and the orange version on the X chromosome inherited from her father (or *vice versa*). Figure 9.5 shows what happens next.

So the tortoiseshell cat ends up with patches of orange and patches of black, depending on the X chromosome that was randomly inactivated in the melanocyte stem cell. The pattern won't change as the cat gets older, it stays the same throughout its life. That tells us that the X inactivation stays the same in the cells that create this coat pattern.

We know that tortoiseshell cats are always female because the gene for the coat colour is only on the X chromosome, not the Y. A male cat only has one X chromosome, so it could have black fur or ginger fur, but never both.

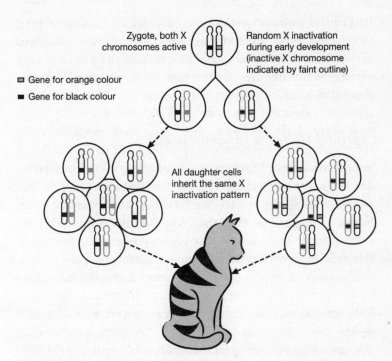

Figure 9.5   In female tortoiseshell cats, the genes for orange and black fur are carried on the X chromosome. Depending on the pattern of X chromosome inactivation in the skin, clonal patches of cells will give rise to discrete patterns of orange and black fur.

Something rather similar happens in a rare human disorder called X-linked hypohidrotic ectodermal dysplasia. This condition is caused by mutations in a gene called *ECTODYSPLASIN-A*, carried on the X chromosome[21]. A male with a mutation in his sole copy of *ECTODYSPLASIN-A* on his single X chromosome has a variety of symptoms, including a total lack of sweat glands. This might sound socially advantageous, but is actually incredibly dangerous. Sweating is one of the major routes by which we lose excess heat, and men with this condition are at serious risk of tissue damage or even death as a result of heat stroke[22].

Females have two copies of the *ECTODYSPLASIN-A* gene, one on each of their X chromosomes. In female carriers of X-linked hypohidrotic ectodermal dysplasia, one X carries a normal copy of the gene, and one a mutated version. There will be random inactivation of one X chromosome in different cells. This means some cells will express a normal copy of *ECTODYSPLASIN-A*. Other cells will randomly shut down the X carrying the normal copy of the gene, and won't be able to express the ECTODYSPLASIN-A protein. Because of the clonal way in which areas of skin develop, just like in the tortoiseshell cat, these women have some patches of skin that express ECTODYSPLASIN-A and some that don't. Where there's no ECTODYSPLASIN-A, the skin can't form sweat glands. As a consequence, these women have patches of skin that can sweat and cool down, and others that can't.

Random X inactivation can significantly influence how females are affected by mutations in genes on the X chromosome. This depends not just on the type of gene that is mutated but also on the tissues that express and require the protein encoded by that gene. The disease called mucopolysaccharidosis II (MPSII) is caused by mutations in the *LYSOSOMAL IDURONATE-2-SULFATASE* gene, on the X chromosome. Boys with this mutation on their single X chromosome are unable to break down certain large molecules and these build up to toxic levels in cells. The main symptoms include airway infections, short stature and enlargement of the spleen and liver. Severely affected boys also suffer mental retardation, and may die in their teenage years.

Females with a mutation in the same gene are usually perfectly healthy. LYSOSOMAL IDURONATE-2-SULFATASE protein is usually secreted out of the cell that makes it and taken up by neighbouring cells. In this situation it doesn't matter too much which X chromosome has been mutated in a specific cell. For every cell that has inactivated the X carrying the normal version of the gene, there is likely to be another cell nearby which inactivated the other X chromosome and is secreting the protein. This way, all cells end up with sufficient LYSOSOMAL IDURONATE-2-SULFATASE protein, whether they produce it themselves or not[23].

Duchenne muscular dystrophy is a severe muscle wasting disease caused by mutations in the X-linked *DYSTROPHIN* gene. This is a large gene that encodes a big protein which acts as an essential shock absorber in muscle fibres. Boys carrying certain mutations in *DYSTROPHIN* suffer major muscle loss that usually results in death in the teenage years. Females with the same mutation are usually symptom-free. The reason for this is that muscle has a very unusual structure. It is called a syncytial tissue, which means that lots of individual cells fuse and operate almost like one giant cell, but with lots of discrete nuclei. This is why most females with a *DYSTROPHIN* mutation are symptom-free. There is enough normal DYSTROPHIN protein encoded by the nuclei that switched off the mutated *DYSTROPHIN* gene to keep this syncytial tissue functioning healthily[24].

There are occasional cases where this system breaks down. There was a case of female monozygotic twins where one twin was severely affected by Duchenne muscular dystrophy and the other was healthy[25]. In the affected twin, the X inactivation had become skewed. Early in tissue differentiation the majority of her cells that would give rise to muscle happened, by ill chance, to switch off the X chromosome carrying the normal copy of the *DYSTROPHIN* gene. Thus, most of the muscle tissue in this woman only expressed the mutated version of DYSTROPHIN, and she developed severe muscle wasting. This could be considered the ultimate demonstration of the power of a random epigenetic event. Two identical individuals, each with two apparently identical X chromosomes, had a completely discordant phenotype, because of a shift in the epigenetic balance of power.

Sometimes, however, it is essential that *individual cells* express the correct amount of a protein. You may have noticed in Chapter 4 that Rett syndrome only affected girls. One might hypothesise that boys are somehow very resistant to the effects of the *MeCP2* mutation, but actually the opposite is true. *MeCP2* is carried on the X chromosome so a male foetus that inherits a Rett syndrome mutation in this gene has no means of expressing normal MeCP2 protein. A complete lack of normal MeCP2 expression is generally lethal in early development, and that's why very few boys are born with Rett syndrome. Girls have two copies of the *MeCP2* gene, one on each X chromosome. In any given cell, there is a 50 per cent chance that the cell will inactivate the X that carries the unmutated *MeCP2* gene and that the cell will not express normal MeCP2 protein. Although a female foetus can develop, there are ultimately major effects on normal post-natal brain development and function when a substantial number of neurons lack MeCP2 protein.

One, two, many

There are other issues that can develop around the X chromosome. One of the questions we need to answer about X inactivation, is how good mammalian cells are at counting. In 2004 Peter Gordon of Columbia University in New York reported on his studies on the Piraha tribe in an isolated region of Brazil. This tribe had numbers for one and two. Everything beyond two was described by a word roughly equating to 'many'[26]. Are our cells the same, or can they count above two? If a nucleus contains more than two X chromosomes, can the X inactivation machinery recognise this, and deal with the consequences? Various studies have shown that it can. Essentially, no matter how many X chromosomes (or more strictly speaking X Inactivation Centres) are present in a nucleus, the cell can count them and then inactivate multiple X chromosomes until there is only one remaining active.

This is the reason why abnormal numbers of X chromosomes are relatively frequent in humans, in contrast to abnormalities in the number of autosomes. The commonest examples are shown in Table 9.1.

Name of syndrome	Karyotype i.e. set of chromosomes	Gender	Frequency of live births (known cases, probably under-estimated)	Common symptoms
Turner	45,X	Female	1 in 2,500	Short stature Infertility Webbed neck Kidney abnormalities
Trisomy X	47,XXX	Female	1 in 1,000	Tall stature Infertility Unusual facial features Poor muscle tone
Klinefelter's	47,XXY	Male	1 in 1,000	Lanky build or rounded body type Infertility Language deficits

Table 9.1 Summary of the major characteristics of the commonest abnormalities in sex chromosome number in humans.

The infertility that is a feature of all these disorders is in part due to problems when creating eggs or sperm, where it's important that chromosomes line up in their pairs. If there is an uneven total number of sex chromosomes this stage goes wrong and formation of gametes is severely compromised.

Leaving aside the infertility, there are two obvious conclusions we can draw from this table. The first is that the phenotypes are all relatively mild compared with, for example, trisomy of chromosome 21 (Down's syndrome). This suggests that cells can tolerate having too many or too few copies of the X chromosome much better than having extra copies of an autosome. But the other obvious conclusion is that an abnormal number of X chromosomes does indeed have some effects on phenotype.

Why should this be? After all, X inactivation ensures that no matter how many X chromosomes are present, all bar one get inactivated early in development. But if this was the end of the story there would be no difference in phenotype between 45, X females compared with 47, XXX females or with the normal 46,

XX female constitution. Similarly, males with the normal 46, XY karyotype should be phenotypically identical to males with the 47, XXY karyotype. In all of these cases there should be only one active X chromosome in the cells.

One thought as to why people with these karyotypes were clinically different was that maybe X inactivation is a bit inefficient in some cells, but this doesn't seem to be the case. X inactivation is established very early in development and is the most stable of all epigenetic processes. An alternative explanation was required.

The answer has its origin about 150 million years ago, when the XY system of sex determination in placental mammals first developed. The X and Y chromosomes are probably descendants of autosomes. The Y chromosome has changed dramatically, the X chromosome much less so[27]. However, both retain shadows of their autosomal past. There are regions on both the X and the Y called pseudoautosomal regions. The genes in these regions are found on both the X and the Y chromosome, just in the same way as pairs of autosomes have the same genes in the same positions, one inherited from each parent.

When an X chromosome inactivates, these pseudoautosomal regions are spared. This means that, unlike most X-linked genes, those in the pseudoautosomal regions don't get switched off. Consequently, normal cells potentially express two copies of these genes in all cells. The two copies are expressed either from the two X chromosomes in a normal female or from the X and the Y in a normal male.

But in Turner's syndrome, the affected female only has one X chromosome, so she expresses only one copy of the genes in the pseudoautosomal region, half as much as normal. In Trisomy X, on the other hand, there are three copies of the genes in the pseudoautosomal regions. As a result, the cells in an affected region will produce proteins from these genes at 50 per cent above the normal level.

One of the genes in the X chromosome pseudoautosomal regions is called *SHOX*. Patients with mutations in this gene have short stature. It is likely that this is also why patients with Turner's syndrome tend to be short – they don't produce enough SHOX protein in their cells. By contrast, patients with Trisomy X are likely to produce 50 per cent more SHOX protein than normal, which is probably why they tend to be tall[28].

It's not just humans who have trisomies of the sex chromosomes. One day you may be happily amazing your friends with your confident statement that their tortoiseshell cat is female when they deflate you by telling you that their pet has been sexed by the vet and is actually a Tom. At this point, smile smugly and then say 'Oh, in that case he's karyotypically abnormal. He has an XXY karyotype, rather than XY'. And if you're feeling particularly mean, you can tell them that Tom is infertile. That should shut them up.

Chapter 10

The Message is Not the Medium

Science commits suicide when it adopts a creed.
Thomas Henry Huxley

One of the most influential books on the philosophy of science is Thomas Kuhn's *The Structure of Scientific Revolutions*, published in 1962. One of the claims in Kuhn's book is that science does not proceed in an orderly, linear and polite fashion, with all new findings viewed in a completely unbiased way. Instead, there is a prevailing theory which dominates a field. When new conflicting data are generated, the theory doesn't immediately topple. It may get tweaked slightly, but scientists can and often do continue to believe in a theory long after there is sufficient evidence to discount it.

We can visualise the theory as a shed, and the new conflicting piece of data as an oddly shaped bit of builder's rubble that has been cemented onto the roof. Now, we can probably continue cementing bits of rubble onto the roof for quite some time, but eventually there will come a point when the shed collapses under the sheer weight of odd bits of masonry. In science, this is when a new theory develops, and all those bits of masonry are used to build the foundations of a new shed.

Kuhn described this collapse-and-rebuild as the paradigm shift, introducing the phrase that has now become such a cliché in the high-end media world. The paradigm shift isn't just based on pure rationality. It involves emotional and sociological changes in the psyches of the upholders of the prevailing theory. Many years before Thomas Kuhn's book, the great German scientist Max Planck, winner of the 1918 Nobel Prize for Physics, put this rather more succinctly when he wrote that, 'Scientific theories don't change because old scientists change their minds; they change because old scientists die[1].'

We are in the middle of just such a paradigm shift in biology.

In 1965, the Nobel Prize in Physiology or Medicine was awarded to François Jacob, André Lwoff and Jacques Monod 'for their discoveries concerning genetic control of enzyme and virus synthesis'. Included in this work was the discovery of messenger RNA (mRNA), which we first met in Chapter 3. mRNA is the relatively short-lived molecule that transfers the information from our chromosomal DNA and acts as the intermediate template for the production of proteins.

We've known for many years that there are some other classes of RNA in our cells, specifically molecules called transfer RNA (tRNA) and ribosomal RNA (rRNA). tRNAs are small RNA molecules that can hold a specific amino acid at one end. When an mRNA molecule is read to form a protein, a tRNA carries its amino acid to the correct place on the growing protein chain. This takes place at large structures in the cytoplasm of a cell called ribosomes. The ribosomal RNA is a major component of ribosomes, where it acts like a giant scaffold to hold various other RNA and protein molecules in position. The world of RNA therefore seemed quite straightforward. There were structural RNAs (the tRNA and rRNA) and there was messenger RNA.

For decades, the stars of the molecular biology catwalk were DNA (the underlying code) and proteins (the functional, can-do molecules of the cell). RNA was relegated to being a relatively uninteresting intermediate molecule, carrying information from a blueprint to the workers on the factory floor.

Everyone working in molecular biology accepts that proteins are immensely important. They carry out a huge range of functions that enable life to happen. Therefore, the genes that encode proteins are also incredibly important. Even small changes to these protein-coding genes can result in devastating effects, such as the mutations that cause haemophilia or cystic fibrosis.

But this world view has potentially left the scientific community a bit blinkered. The fact that proteins, and therefore by extension protein-coding genes, are vitally important should not imply that everything else in the genome is unimportant. Yet this is the theoretical construct that has applied for decades now. That's actually quite odd, because we've had access for many years to data that show that proteins can't be the whole story.

Why we don't throw away our junk

Scientists have recognised for some time that the blueprint is edited by cells before it is delivered to the workers. This is because of introns, which we met in Chapter 3. They are the sequences that are copied from DNA into mRNA, but then spliced out before the message is translated into a protein sequence by the ribosomes. Introns were first identified in 1975[2] and the Nobel Prize for their discovery was awarded to Richard Roberts and Phillip Sharp in 1993.

Back in the 1970s scientists compared simple one-celled organisms and complex creatures like humans. The amount of DNA in their cells seemed surprisingly similar, considering how dissimilar the organisms were. This suggested that some genomes must contain a lot of DNA that isn't really used for anything, and led to the concept of 'junk DNA'[3] – chromosome sequences that don't do anything useful, because they don't code for proteins. At around the same time a number of labs showed that large amounts of the mammalian genome contain DNA sequences that seem to be repeated over and over again, and don't code for proteins (repetitive DNA). Because they don't code for protein, it was assumed they weren't contributing anything to the cell's functions. They just appeared to be along for the ride[4,5]. Francis

Crick and others coined the phrase 'selfish DNA' to describe these regions. These two models, of junk DNA and selfish DNA, have been delightfully described recently as 'the emerging view of the genome as being largely populated by genetic hobos and evolutionary debris[6]'.

We humans are remarkable, with our trillions of cells, our hundreds of cell types, our multitudes of tissues and organs. Let's compare ourselves (a little smugly, perhaps) with a distant relative, a microscopic worm, the nematode *Caenorhabditis elegans*. *C. elegans*, as we usually call it, is only about one millimetre long and lives in soil. It has many of the same organs as higher animals, such as a gut, mouth and gonads. However, it only consists of around 1,000 cells. Remarkably, as *C. elegans* develops, scientists have been able to identify exactly how each of these cells arises.

This tiny worm is a powerful experimental tool, because it provides a roadmap for cell and tissue development. Scientists are able to alter expression of a gene and then plot out with great precision the effects of that mutated gene on normal development. In fact, *C. elegans* has laid the foundation for so many breakthroughs in developmental biology that in 2002 the Nobel Committee awarded the Prize in Physiology or Medicine to Sydney Brenner, Robert Horvitz and John Sulston for their work on this organism.

We can't fault *C. elegans* on grounds of utility, but it is clearly a much less complex organism than our good selves. Why are we so much more sophisticated? Given the importance of proteins in cellular function, the original assumption was that complex organisms like mammals have more protein-coding genes than simple creatures like *C. elegans*. This was a perfectly reasonable hypothesis but it has fallen foul of a phenomenon described by Thomas Henry Huxley. He was Darwin's great champion in the 19th century and it was Huxley who first described 'the slaying of a beautiful hypothesis by an ugly fact'.

As DNA sequencing technologies improved in cost and efficiency, numerous labs throughout the world sequenced the genomes of a number of different organisms. They were able to use various software tools to identify the likely protein-coding genes in these different genomes. What they found was really surprising. There were far fewer protein-coding genes than expected. Before the human genome was decoded, scientists had predicted there would be over 100,000 such genes. We now know the real number is between 20,000 and 25,000 genes[7]. Even more oddly, *C. elegans* contains about 20,200 genes[8], not so very different a number from us.

Not only do we and *C. elegans* have about the same number of genes, these genes tend to code for pretty much the same proteins. By this we mean that if we analyse the sequence of a gene in human cells, we can find a gene of broadly similar sequence in the nematode worm. So the phenotypic differences between worms and humans aren't caused by *Homo sapiens* having more, different or 'better' genes.

Admittedly, more complicated organisms tend to splice their genes in more ways than simpler creatures. Using our CARDIGAN example from Chapter 3 as an analogy once again, *C. elegans* might only be able to make the proteins DIG and DAN whereas mammals would be able to make those two proteins and also CARD, RIGA, CAIN and CARDIGAN.

This certainly would allow humans to generate a much greater repertoire of proteins than the 1mm worm, but it introduces a new problem. How do more complicated organisms regulate their more complicated splicing patterns? This regulation could in theory be controlled solely by proteins, but this in turn has difficulties. The more proteins a cell needs to regulate in a complicated network, the more proteins it needs to do the regulation. Mathematical models have shown that this rapidly leads to a situation where the number of proteins that we need begins to out-strip the number of proteins that we actually possess – clearly a non-starter.

Do we have an alternative? We do, and it's indicated in Figure 10.1.

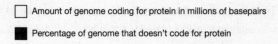

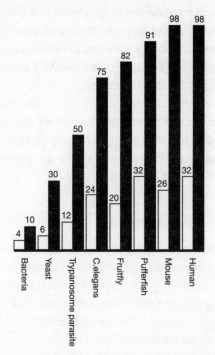

Figure 10.1   This graph demonstrates that the complexity of living organisms scales much better with the percentage of the genome that doesn't code for protein (black columns) than it does with the number of basepairs coding for protein in a genome (white columns). The data are adapted from Mattick, J. (2007), Exp Biol. 210: 1526–1547.

At one extreme we have the bacteria. Bacteria have very small, highly compacted genomes. Their protein-coding genes cover about 4,000,000 base-pairs, which is about 90 per cent of their genome. Bacteria are very simple organisms and fairly rigid in the way they control their gene expression. But things change as we move further up the evolutionary tree.

The protein-coding genes of *C. elegans* cover about 24,000,000 base-pairs, but that only accounts for about 25 per cent of their genome. The remaining 75 per cent doesn't code for protein. By the time we reach humans, the protein-coding regions cover about 32,000,000 base-pairs, but this only represents about 2 per cent of the total genome. There are various ways that we can calculate the protein-coding regions, but they make relatively little difference to the astonishing bottom line. Over 98 per cent of the human genome doesn't code for protein. All but 2 per cent of our genome is 'junk'.

In other words, the numbers of genes, or the sizes of these genes, don't scale with complexity. The only feature of a genome that really seems to get bigger as organisms get more complicated is the section that *doesn't* code for protein.

The tyranny of language

So what are these non-coding regions of the genome doing, and why are they so important? It's when we start to consider this that we begin to notice what a strong effect language and terminology have on human thought processes. These regions are called non-coding, but what we mean is that they don't code for *protein*. This isn't the same as not coding at all.

There is a well-known scientific proverb: absence of evidence is not the same as evidence of absence. For example, in astronomy, once scientists had developed telescopes that could detect infrared radiation, they were able to detect thousands of stars that had never been 'seen' before. The stars had always been there, but we couldn't detect them conclusively until we had an instrument for doing so. A more everyday example might be a mobile phone signal. Such signals are all around us, but we cannot detect them unless we have a mobile phone. In other words, what we find depends very much on how we are looking.

Scientists identify the genes which are expressed in a specific cell type by analysing the RNA molecules. This is done by extracting all the RNA from cells and then analysing it using various different techniques, so that you build a database of all the RNA molecules that are present. When researchers in the 1980s first began investigating which genes were expressed in a given cell type, the techniques available were relatively insensitive. They were also designed to detect only mRNA molecules, as these were the ones that were assumed to be important. These methods tended to be good at detecting highly expressed mRNAs and quite poor at detecting the less well-expressed sequences. Another confounding factor was that the software used to analyse mRNA was set so that it would ignore signals originally generated from repetitive, i.e. 'junk', DNA.

These techniques served us very well for profiling the mRNA that we were already interested in – the mRNA molecules that coded for proteins. But as we have seen, this only represents about 2 per cent of the genome. It wasn't until new detection technologies were coupled with hugely increased computing power that we began to realise that something very interesting was happening in the remaining 98 per cent – the non-coding part of our genome.

With these improved methodologies, the scientific world began to appreciate that there was actually a huge amount of transcription going on in the parts of the genome that didn't code for proteins. Initially this was dismissed as 'transcriptional noise'. It was suggested that there was a baseline murmur of expression from all over the genome, as if these regions of DNA occasionally produced an RNA molecule that got above a detection threshold. The concept was that although we could detect these molecules with our new, more sensitive equipment, they weren't really biologically meaningful.

The phrase 'transcriptional noise' implies a basically random event. However, the patterns of expression of these non-protein-coding RNAs were different for different cell types, which suggested that their transcription was far from random[9]. For example, there was a lot of this expression in the brain. It's now become clear that the patterns of expression are different in different brain regions[10]. This effect is reproducible when the various brain regions are compared from different individuals. This isn't what we would expect if this low-level transcription of RNA was a purely random process.

It is becoming clearer that this transcription from genes that don't code for protein is actually critically important for cellular function. Oddly, however, we remain caught in a linguistic trap of our own making. The RNA that is produced from these regions, the RNA that was previously under our radar, is still called non-coding RNA (ncRNA). It's a sloppy shorthand, because what we really mean is non-*protein*-coding RNA. The ncRNA does, in fact, code for something – it codes for itself, a functional RNA molecule. Unlike mature mRNA, which is an RNA means to a protein end, ncRNAs are themselves the end-points.

Re-defining rubbish

This is the paradigm shift. For at least 40 years molecular biologists and geneticists have focused almost exclusively on the genes that code for proteins, and the proteins themselves. There have been exceptions, but we've just treated these as the odd bits of rubble on the top of the shed. But non-coding RNAs are finally starting to stand firmly alongside proteins as fully functional molecules. Different but equal.

These ncRNAs are found all over the genome. Some come from introns. Originally it was assumed that the spliced-out bits of mRNA from the introns get degraded by cells. It now seems much more likely that at least some (if not all or most) are actually processed to act as functional ncRNAs in their own right. Others overlap genes, frequently transcribed from the opposite strand to the protein-coding mRNA. Yet others are found in regions where there are no protein-coding genes at all.

We met two ncRNAs in the last chapter. These were *Xist* and *Tsix*, the ncRNAs that are required for X inactivation. These are both very long ncRNAs, of several thousand kilobases in length. When *Xist* was first identified, it was only the second known ncRNA. Current estimates suggest there are thousands of such molecules in the cells of higher mammals, with over 30,000 'long' ncRNAs (defined as having a length greater than 200 bases) reported in mice[11]. Long ncRNAs may actually out-number protein-coding mRNAs.

In addition to X inactivation, long ncRNAs also appear to play a critical role in imprinting. Many imprinted regions contain a section that encodes a long ncRNA, which silences the expression of surrounding genes. This is similar to the effect of *Xist*. The protein-coding mRNAs are silenced on the copy of the chromosome which expresses the long ncRNA. For example, there is an ncRNA called *Air*, expressed in the placenta, exclusively from the paternally inherited mouse chromosome 11. Expression of *Air* ncRNA represses the nearby *Igf2r* gene, but only on the same chromosome[12]. This mechanism ensures that *Igf2r* is only expressed from the maternally inherited chromosome.

The *Air* ncRNA gave scientists important insights into how these long ncRNAs repress gene expression. The ncRNA remained localised to a specific region in the cluster of imprinted genes, and acted as a magnet for an epigenetic enzyme called G9a. G9a puts a repressive mark on the histone H3 proteins in the nucleosomes deposited on this region of DNA. This histone modification creates a repressive chromatin environment, which switches off the genes.

This finding was particularly important as it provided some of the first insights into a question that had been puzzling epigeneticists. How do histone modifying enzymes, which put on or remove epigenetic marks, get localised to specific regions of the genome? Histone modifying enzymes can't recognise specific DNA sequences directly, so how do they end up in the right part of the genome?

The patterns of histone modifications are localised to different genes in different cell types, leading to exquisitely well-regulated gene expression. For example, the enzyme known as EZH2 methylates the amino acid called lysine at position 27 on histone H3, but it targets different histone H3 molecules in different cell types. To put it simply, it may methylate histone H3 proteins positioned on gene A in white blood cells but not in neurons. Alternatively, it may methylate histone H3 proteins positioned on gene B in neurons, but not in white blood cells. It's the same enzyme in both cells, but it's being targeted differently.

There is increasing evidence that at least some of the targeting of epigenetic modifications can be explained by interactions with long ncRNAs. Jeannie Lee and her colleagues have recently investigated long ncRNAs that bind to a complex of proteins. The complex is called PRC2 and it generates repressive modifications on histones. PRC2 contains a number of proteins, and the one that interacts with the long ncRNAs is probably EZH2. The researchers found that the PRC2 complex bound to literally thousands of different long ncRNA molecules in embryonic stem cells from mice[13]. These long ncRNAs may act as bait. They can stay tethered to the specific region of the genome where they are produced, and then attract repressive enzymes to shut off gene expression. This happens because the repressive enzyme complexes contain proteins like EZH2 that are capable of binding to RNA.

Scientists love to build theories, and in some ways a nice one was shaping up around long ncRNAs. It seemed that they bind to the region from which they are transcribed, and repress gene expression on that same chromosome. But if we go back to our analogy from the start of this chapter, we'd have to say that it's now becoming clear we have built a pretty small shed and already cemented quite a bit of rubble to the roof.

There's an amazing family of genes, called *HOX* genes. When they're mutated in fruit flies (*Drosophila melanogaster*) the results are incredible phenotypes, such as legs growing out of the head[14]. There's a long ncRNA known as *HOTAIR*, which regulates a region of genes called the *HOX-D* cluster. Just like the long ncRNAs investigated by Jeannie Lee, *HOTAIR* binds the PRC2 complex and creates a chromatin region which is marked with repressive histone modifications. But *HOTAIR* is not transcribed from the *HOX-D* position on chromosome 12. Instead it is encoded at a different cluster of genes called *HOX-C* on chromosome 2[15]. No-one knows how or why *HOTAIR* binds at the HOX-D position.

There's a related mystery around the best studied of all long ncRNAs, *Xist*. *Xist* ncRNA spreads out along almost the entire inactive X chromosome but we really don't know how. Chromosomes don't normally become smothered with RNA molecules. There's no obvious reason why *Xist* RNA should be able to bind like this, but we know it's nothing to do with the sequence of the chromosome. The experiments described in the last chapter, where *Xist* could inactivate an entire autosome as long as it contained an X inactivation centre, showed that *Xist* just keeps on travelling once it's on a chromosome. Scientists are basically still completely baffled about these fundamental characteristics of this best-studied of all ncRNAs.

Here's another surprising thing. Until very recently, all long ncRNAs were thought to repress gene expression. In 2010, Professor Ramin Shiekhattar at the Wistar Institute in Philadelphia identified over 3,000 long ncRNAs in a number of human cell types. These long ncRNAs showed different expression patterns in different human cell types, suggesting they had specific roles. Professor Shiekhattar and his colleagues tested a small number of the long ncRNAs to try to determine their functions. They used well-established experimental methods to knock down expression of their test ncRNAs and then analysed expression of their neighbouring genes. The predicted outcome, and the actual results, are shown in Figure 10.2.

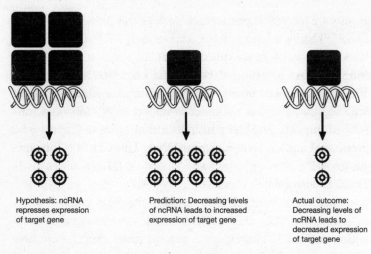

Hypothesis: ncRNA
represses expression
of target gene

Prediction: Decreasing levels
of ncRNA leads to increased
expression of target gene

Actual outcome:
Decreasing levels of
ncRNA leads to
decreased expression
of target gene

Figure 10.2   ncRNAs were thought to repress expression of target genes. If this hypothesis were correct, then decreasing the expression of a specific ncRNA should result in more expression of the target gene, as the repression diminishes. This is shown in the middle panel. However, it is now becoming clear that a large number of ncRNAs actually drive *up* expression of their target genes. This has been shown by cases in which experimentally decreasing the expression of an ncRNA has the effect shown in the right hand side of this figure.

Twelve ncRNAs were tested, and in seven cases the scientists found the result shown in the right-hand panel of Figure 10.2. This was contrary to expectations, because it suggests that about 50 per cent of long ncRNAs may actually increase expression of neighbouring genes, not decrease it[16].

Rather pithily, the authors of the paper stated, 'The precise mechanism by which our ncRNAs function to enhance gene expression is not known.' It's a statement that is very hard to argue with. It has considerable merit as it makes clear that we currently have no idea how this is happening. Ramin Shiekhattar's work does demonstrate rather convincingly that there is a lot we don't understand about long ncRNAs, and that we should be wary of creating new dogma too quickly.

Small is beautiful

We should also be wary of assuming that size is everything and that big is best. The long ncRNAs clearly have major importance in cell function, but there is another equally importance class of ncRNAs that also has a significant impact in the cell. The ncRNAs in this class are short (usually 20–24 bases in length), and they target mRNA molecules, not DNA. This was first shown in our favourite worm, *C. elegans*.

As we have already discussed, *C. elegans* is a very useful model system because we know exactly how every cell should normally develop. The timing and sequence of the different stages is very tightly regulated. One of the key regulators is a protein called LIN-14. The *LIN-14* gene is highly expressed (a lot of LIN-14 protein is produced) during the very early embryo stages, but is down-regulated as the worms move from larval stage 1 to larval stage 2. If the *LIN-14* gene is mutated the worm gets the timing of the different stages wrong. If LIN-14 protein stays on for too long the worm starts to repeat early developmental stages. If LIN-14 protein is lost too early the worm moves into later larval stages prematurely. Either way, the worm gets very messed up, and normal adult structures don't develop.

In 1993 two labs working independently showed how *LIN-14* expression was controlled[17,18]. Unexpectedly, the key event was binding of a small ncRNA to the *LIN-14* mRNA molecule. This is shown in Figure 10.3. It is an example of post-transcriptional gene silencing, where an mRNA is produced but is prevented from generating a protein. This is a very different way of controlling gene expression from that used by the long ncRNAs.

The importance of this work is that it laid the foundation for a whole new model for the regulation of gene expression. Small ncRNAs are now known to be a mechanism used by organisms throughout the plant and animal kingdoms to control gene expression. There are various different types of

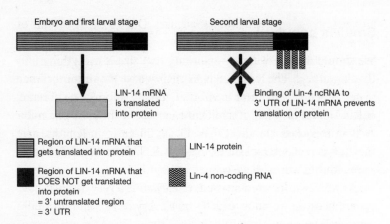

Embryo and first larval stage | Second larval stage

LIN-14 mRNA
is translated
into protein

Binding of Lin-4 ncRNA to
3' UTR of LIN-14 mRNA prevents
translation of protein

Region of LIN-14 mRNA that
gets translated into protein

LIN-14 protein

Region of LIN-14 mRNA that
DOES NOT get translated
into protein
= 3' untranslated region
= 3' UTR

Lin-4 non-coding RNA

Figure 10.3 Schematic to demonstrate how expression of microRNAs at specific developmental stages can radically alter expression of a target gene.

small ncRNAs, but we'll concentrate mainly on the microRNAs (miRNAs).

At least 1,000 different miRNAs have been identified in mammalian cells. miRNAs are about 21 nucleotides (bases) in length (sometimes slightly smaller or longer) and most of them seem to act as post-transcriptional regulators of gene expression. They don't stop production of an mRNA, instead they regulate how that mRNA behaves. Typically, they do this by binding to the 3' untranslated region (3' UTR) of an mRNA molecule. This region is shown in Figure 10.3. It's present in the mature mRNA, but it doesn't code for any amino acids.

When genomic DNA is copied to make mRNA, the original transcript tends to be very long because it contains both exons (which code for amino acids) and introns (which do not). As we saw in Chapter 3, introns are removed during splicing to create an mRNA which codes for protein. But the Chapter 3 description passed over something. There are stretches of RNA at the beginning (known as 5' UTR) and the end (3' UTR) which don't code for amino acids, but don't get spliced out like introns either. Instead, these non-coding regions are retained on the mature mRNA and act as regulatory sequences. One of the functions of the 3′ UTR in particular is to bind regulatory molecules, including miRNAs.

How does a miRNA bind to an mRNA and what happens when it does? The miRNA and the 3′ UTR of the mRNA only interact if they recognise each other. This uses base-pairing, quite similar to that in double stranded DNA. G can bind C, A can bind U (in RNA, T is replaced by U). Although miRNAs are usually 21 bases in length, they don't have to match the mRNA over the entire 21 nucleotides. The key region is positions 2 to 8 on the miRNA.

Sometimes the match from 2 to 8 is not perfect, but it's still close enough for the two molecules to pair up. In these cases, binding of the miRNA prevents translation of the mRNA into protein (this is what happened in the case shown in Figure 10.3). If, however, the match is perfect, the binding of miRNA to mRNA triggers destruction of the mRNA, by enzymes that attach to the miRNA[19]. It's not yet clear if positions 9 to 21 on the miRNAs also influence in a less direct way how these small molecules are targeted, or what the consequences of their targeting are. One thing we do know, however, is that a single miRNA can regulate more than one mRNA molecule. We saw in Chapter 3 how one gene could encode lots of different protein molecules, by altering the way in which messenger RNA is spliced. A single miRNA can influence many of these differently spliced versions simultaneously. Alternatively, a single miRNA can also influence quite unrelated proteins that are encoded by different genes but have similar 3′ UTR sequences.

This can make it very difficult to unravel exactly what a miRNA is doing in a cell, as the effects will vary depending on the cell type and the other genes (protein-coding and non-protein-coding) that the cell is expressing at any one time. That can be important experimentally, but also has significant consequences for normal health and disease. In conditions where there are an abnormal number of chromosomes, for example, it won't just be protein-coding genes that change in number. There will also be abnormal production of ncRNAs (large and small). Because miRNAs in particular can regulate lots of other genes, the effects of disrupting miRNA copy numbers may be very extensive.

Room for manoeuvre

The fact that 98 per cent of the human genome does not code for protein suggests that there has been a huge evolutionary investment in the development of complicated ncRNA-mediated regulatory processes. Some authors have even gone so far as to speculate that ncRNAs are the genetic features that have underpinned the development of *Homo sapiens'* greatest distinguishing feature – our higher thought processes[20].

The chimpanzee is our closest relative and its genome was published in 2005[21]. There isn't one simple, meaningful average figure that we can give to express how similar the human and chimp genomes are. The statistics are actually very complicated, because you have to take into account that different genomic regions (for example repetitive sections versus single copy protein-coding gene regions) affect the statistics differently. However, there are two things we can say quite firmly. One is that human and chimp proteins are incredibly similar. About a third of all proteins are exactly the same between us and our knuckle-dragging cousins, and the rest differ only by one or two amino acids. Another thing we have in common is that over 98 per cent of our genomes don't code for protein. This suggests that both species use ncRNAs to create complex regulatory networks which govern gene and protein expression. But there is a particular difference which may be very important between chimps and humans. This lies in how ncRNA is treated in the cells of the two species.

It's all to do with a process called editing. It seems that human cells just can't leave well-enough alone, particularly when it comes to ncRNA[22]. Once an ncRNA has been produced, human cells use various mechanisms to modify it yet further. In particular, they will often change the base A to one called I (inosine). Base A can bind to T in DNA, or U in RNA. But base I can pair with A, C or G. This alters the sequences to which an ncRNA can bind and hence regulate.

We humans, more than any other species, edit our ncRNA molecules to a remarkable degree. Not even other primates carry out this reaction as well as we do[23]. We also edit particularly extensively in the brain. This makes editing of ncRNA an attractive candidate process to explain why we are mentally so much more sophisticated than our primate relatives, even though we share so much of our DNA template in common.

In some ways, this is the beauty of ncRNAs. They create a relatively safe method for organisms to use to alter various aspects of cellular regulation. Evolution has probably favoured this mechanism because it is simply too risky to try to improve function by changing proteins. Proteins, you see, are the Mary Poppins of the cell. They are 'practically perfect in every way'.

Hammers always look pretty similar. Some may be big, some may be small, but in terms of basic design, there's not much you can change that would make a hammer much better. It's the same with proteins. The proteins in our bodies have evolved over billions of years. Let's take just one example. Haemoglobin is the pigment that transports oxygen around our bodies, in the red blood cells. It's beautifully adept at picking up oxygen in the lungs and releasing it where it's needed in the tissues. Nobody working in a lab has been able to create an altered version of haemoglobin that does a better job than the natural protein.

Creating a haemoglobin molecule that's worse than normal is surprisingly easy to do, unfortunately. In fact, that's what happens in disorders like sickle cell disease, where mutations create poor haemoglobin proteins. A similar situation is true for most proteins. So, unless environmental conditions change dramatically, most alterations to a protein turn out to be a bad thing. Most proteins are as good as they're going to get.

So how has evolution solved the problem of creating ever more complex and sophisticated organisms? Basically, by altering the *regulation* of proteins, rather than altering the proteins themselves. This is what can be achieved using complicated networks of ncRNA molecules to influence how, when and to what degree specific proteins are expressed – and there is evidence to show this actually happens.

miRNAs play major roles in control of pluripotency and control of cellular differentiation. ES cells can be encouraged to differentiate into other cell types by changing the culture conditions in which they're grown. When they begin to differentiate, it's essential that ES cells switch off the gene expression pathways that normally allow them to keep producing additional ES cells (self-renewal). There is a miRNA family called *let-7* which is essential for this switch-off process[24].

One of the mechanisms the *let-7* family uses is the down-regulation of a protein called Lin28. This implies that Lin28 is a pro-pluripotency protein. It's therefore not that surprising to discover that Lin28 can act as a Yamanaka factor. Over-expression of Lin28 protein in somatic cells increases the chances of reprogramming them to iPS cells[25].

Conversely, there are other miRNA families that help ES cells to stay pluripotent and self-renewing. Unlike *let-7*, these miRNAs promote the pluripotent state. In ES cells, the key pluripotency factors such as Oct4 and Sox2 are bound to the promoters of these miRNAs, activating their expression. As the ES cells start to differentiate, these factors fall off the miRNA promoters, and stop driving their expression[26]. Just like the Lin28 protein, these miRNAs also improve reprogramming of somatic cells into iPS cells[27].

When we compare stem cells with their differentiated descendants, we find that they express very different populations of mRNA molecules. This seems reasonable, as the stem and differentiated cells express different proteins. But some mRNAs can take a long time to break down in a cell. This means that when a stem cell starts to differentiate, there will be a period when it still contains many of the stem cell mRNAs. Happily, when the stem cell starts differentiating, it switches on a new set of miRNAs. These target the residual stem cell mRNAs and accelerate their destruction. This rapid degradation of the pre-existing mRNAs ensures that the cell moves into a differentiated state as quickly and irreversibly as possible[28].

This is an important safety feature. It's not good for cells to retain inappropriate stem cell characteristics – it increases the chance they will move down a cancer cell pathway. This mechanism is used even more dramatically in species where embryonic development is very rapid, such as fruit flies or zebrafish. In these species this process ensures that maternally-inherited mRNA transcripts supplied by the egg are rapidly degraded as the fertilised egg turns into a pluripotent zygote[29].

miRNAs are also vital for that all-important phase in imprinting control, the formation of primordial germ cells. A key stage in creation of primordial germ cells is the activation of the Blimp1 protein that we met in Chapter 8. Blimp1 expression is controlled by a complex interplay between Lin28 and let-7 activity[30]. Blimp1 also regulates an enzyme that methylates histones, and a class of proteins known as PIWI proteins. PIWI proteins in turn bind to another type of small ncRNAs known as PIWI RNAs[31]. PIWI ncRNAs and proteins don't seem to play much of a role in the somatic cells but are required for generation of the male germline[32]. PIWI actually stands for *P* element-*i*nduced *wi*mpy testis. If the PIWI ncRNAs and PIWI proteins don't interact properly, the testes in a male foetus don't form normally.

We are finding more and more instances of cross-talk and interactions between ncRNAs and epigenetic events. Remember that the genetic interlopers, the retrotransposons, are normally methylated in the germline, to prevent their activation. The PIWI pathway is involved in targeting this DNA methylation[33,34]. A substantial number of epigenetic proteins are able to interact with

RNA. Binding of non-coding RNAs to the genome may act as the general mechanism by which epigenetic modifications are targeted to the correct chromatin region in a specific cell type[35].

ncRNAs have recently been implicated in Lamarckian transmission of inherited characteristics. In one example, fertilised mouse eggs were injected with a miRNA which targeted a key gene involved in growth of heart tissue. The mice which developed from these eggs had enlarged hearts (cardiac hypertrophy) suggesting that the early injection of the miRNA disturbed the normal developmental processes. Remarkably, the offspring of these mice also had a high frequency of cardiac hypertrophy. This was apparently because the abnormal expression of the miRNA was recreated during generation of sperm in these mice. There was no change in the DNA code of the mice, so this was a clear case of a miRNA driving epigenetic inheritance[36].

Murphy's Law (if something can go wrong, it usually will)

But if ncRNAs are so important for cellular function, surely we would expect to find that sometimes diseases are caused by problems with them. Shouldn't there be lots of examples where defects in production or expression of ncRNAs lead to clinical disorders, aside from the imprinting or X inactivation conditions? Well, yes and no. Because these ncRNAs are predominantly regulatory molecules, acting in networks that are rich in compensatory mechanisms, defects may only have relatively subtle impacts. The problem this creates experimentally is that most genetic screens are good at detecting the major phenotypes caused by mutations in proteins, but may not be so useful for more subtle effects.

There is a small ncRNA called BC1 which is expressed in specific neurons in mice. When researchers at the University of Munster in Germany deleted this ncRNA, the mice seemed fine. But then the scientists moved the mutant animals from the very controlled laboratory setting into a more natural environment. Under these conditions, it became clear that the mutants were not the same as normal mice. They were reluctant to explore their surroundings and were anxious[37]. If they had simply been left in their cages, we would never have appreciated that loss of the BC1 ncRNA actually had a quite pronounced effect on behaviour. A clear case of what we see being dependent on how we look.

The impact of ncRNAs in clinical conditions is starting to come into focus, at least for a few examples. There is a breed of sheep called a Texel, and the kindest description would be that it's chunky. The Texel is well known for having a lot of muscle, which is a good thing in an animal that's being bred to be eaten. The muscularity of the breed has been shown to be at least partially due to a change in a miRNA binding site in the 3′ UTR of a specific gene. The protein coded for by this gene is called myostatin, and it normally slows down muscle growth[38]. The impact of the single base change is summarised in Figure 10.4. The final size of the Texel sheep has been exaggerated for clarity.

Tourette's syndrome is a neurodevelopmental disorder where the patient frequently suffers from involuntary convulsive movements (tics) which in some cases are associated with involuntary swearing. Two unrelated individuals with this disorder were shown to have the same single base change in the 3′ UTR of a gene called *SLITRK1*[39]. *SLITRK1* appears to be required for neuronal development. The base change in the Tourette's patients introduced a binding site for a short ncRNA called miR-189. This suggests that *SLITRK1* expression may be abnormally down-regulated via such binding, at critical points in development. This alteration is only present in a few cases of Tourette's but raises the tantalising suggestion that mis-regulation of miRNA binding sites in other neuronal genes may be involved in other patients.

Earlier in this chapter we encountered the theory that ncRNAs may have been vitally important for the development of increased brain complexity and sophistication in humans. If that is the case,

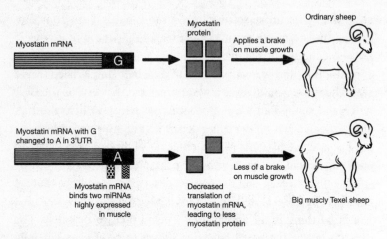

Figure 10.4   A single base change which is in a part of the myostatin gene that does not code for protein nevertheless has a dramatic impact on the phenotype in the Texel sheep breed. The presence of an A base instead of a G in the myostatin mRNA leads to binding of two specific miRNAs. This alters myostatin expression, resulting in sheep with very pronounced muscle growth.

we might predict that the brain would be particularly susceptible to defects in ncRNA activity and function. Indeed, the Tourette's cases in the previous paragraph give an intriguing glimpse of such a scenario.

There is a condition in humans called DiGeorge syndrome in which a region of about 3,000,000 bases has been lost from one of the two copies of chromosome 22[40]. This region contains more than 25 genes. It's probably not surprising that many different organ systems may be affected in patients with this condition, including genito-urinary, cardiovascular and skeletal. Forty per cent of DiGeorge patients suffer seizures and 25 per cent of adults with this condition develop schizophrenia. Mild to moderate mental retardation is also common. Different genes in the 3,000,000 base-pair region probably contribute to different aspects of the disorder. One of the genes is called *DGCR8* and the DGCR8 protein is essential for the normal production of miRNAs. Genetically modified mice have been created with just one functional copy of *Dgcr8*. These mice develop cognitive problems, especially in learning and spatial processing[41]. This supports the idea that miRNA production may be important in neurological function.

We know that ncRNAs are important in the control of cellular pluripotency and cellular differentiation. It's not much of a leap from that to hypothesise that miRNAs may be important in cancer. Cancer is classically a disease in which cells can keep proliferating. This has parallels with stem cells. Additionally, in cancer, the tumours often look relatively undifferentiated and disorganised under the microscope. This is in contrast to the fully differentiated and well-organised appearance of normal, healthy tissues. There is now a strong body of evidence that ncRNAs play a role in cancer. This role may involve either loss of selected miRNAs or over-expression of other miRNAs, as shown in Figure 10.5.

Chronic lymphocytic leukaemia is the commonest human leukaemia. Approximately 70 per cent of cases of this type of cancer[42] have lost the ncRNAs called *miR-15a* and *miR-16-1*. Cancer is a multi-step disease and a lot of things need to go wrong in an individual cell before it becomes cancerous. The fact that so many cases of this type of leukaemia, the most common human leukaemia, lacked these particular miRNAs suggested that loss of these sequences happened early in the development of the disease.

An example of the alternative mechanism – over-expression of miRNAs in cancer – is the case of the *miR-17-92* cluster. This cluster is over-expressed in a range of cancers[43]. In fact, a considerable number of reports have now been published on abnormal expression of miRNAs in cancer[44]. In addition, a gene called *TARBP2* is mutated in some inherited cancer conditions[45]. The TARBP2 protein is involved in normal processing of miRNAs. This strengthens the case for a role of miRNAs in the initiation and development of certain human cancers.

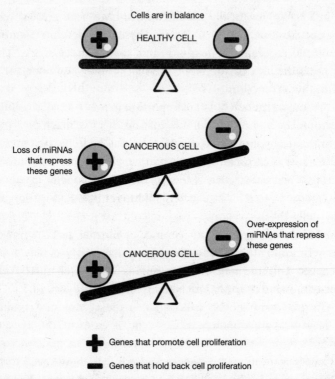

Figure 10.5   Decreased levels of certain types of microRNAs, or increased levels of others, may each ultimately have the same disruptive effect on gene expression. The end result may be increased expression of genes that drive cells into a highly proliferative state, increasing the likelihood of cancer development.

Hope or hype?

Given the increasing amounts of data suggesting a major role for miRNAs in cancer, it isn't surprising that scientists began to get excited about the possibilities of using these molecules to treat cancer. The idea would be to replace 'missing' miRNAs or to inhibit ones that were over-expressed. The hope was that this could be achieved by dosing cancer patients with the miRNAs, or artificial variants of them. This could also have applications in other diseases where miRNA expression may have become abnormal.

Big pharmaceutical companies are certainly investing heavily in this area. Sanofi-Aventis and GlaxoSmithKline have each formed multi-million dollar collaborations with a company called Regulus Therapeutics in San Diego. They are exploring the development of miRNA replacements or inhibitors, to use in the treatment of diseases ranging from cancer to auto-immune disorders.

There are molecules very like miRNAs called siRNAs (small interfering RNAs). They use much the same processes as miRNA molecules to repress gene expression, especially degradation of mRNA. siRNAs have been used as tools very extensively in research, as they can be administered to cells in culture to switch off a gene for experimental investigations. In 2006, the scientists who first developed this technology, Andrew Fire and Craig Mello, were awarded the Nobel Prize for Physiology or Medicine.

Pharmaceutical companies became very interested in using siRNAs as potential new drugs. Theoretically, siRNA molecules could be used to knock down expression of any protein that was believed to be harmful in a disease. In the same year that Fire and Mello were awarded their Nobel Prize, the giant pharmaceutical company Merck paid over one billion US dollars for a siRNA company in California called Sirna Therapeutics. Other large pharmaceutical companies have also invested heavily.

But in 2010 a bit of a chill breeze began to drift through the pharmaceutical industry. Roche, the giant Swiss company, announced that it was stopping its siRNA programmes, despite having spent more than $500 million on them over three years. Its neighbouring Swiss corporation, Novartis, pulled out of a collaboration with a siRNA company called Alnylam in Massachusetts. There are still plenty of other companies who have stayed in this particular game, but it would probably be fair to say there's a bit more nervousness around this technology than in the past.

One of the major problems with using this kind of approach therapeutically may sound rather mundane. Nucleic acids, such as DNA and RNA, are just difficult to turn into good drugs. Most good existing drugs – ibuprofen, Viagra, anti-histamines – have certain characteristics in common. You can swallow them, they get across your gut wall, they get distributed around your body, they don't get destroyed too quickly by your liver, they get taken up by cells, and they work their effects on the molecules in or on the cells. Those all sound like really simple things, but they're often the most difficult things to get right when developing a new drug. Companies will spend tens of millions of dollars – at least – getting this bit right, and it is still a surprisingly hit-and-miss process.

It's so much worse when trying to create drugs around nucleic acids. This is partly because of their size. An average siRNA molecule is over 50 times larger than a drug like ibuprofen. When creating drugs (especially ones to be taken orally rather than injected) the general rule is, the smaller the better. The larger a drug is, the greater the problems with getting high enough doses into patients, and keeping them in the body for long enough. This may be why a company like Roche has decided it can spend its money more effectively elsewhere. This doesn't mean that siRNA won't ever work in the treatment of illnesses, it's just quite high risk as a business venture. miRNA essentially faces all the same problems, because the nucleic acids are so similar for both approaches.

Luckily, there is usually more than one way to treat a cat and in the next chapter, we'll see how drugs targeting epigenetic enzymes are already treating patients with severe cancer conditions.

Chapter 11

Fighting the Enemy Within

The most exciting phrase to hear in science, the one that heralds new discoveries, is not 'Eureka!' (I found it!) but 'That's funny ...'
Isaac Asimov

There are multiple instances in science of a relatively chance event leading to a wonderful breakthrough. Probably the most famous example is Alexander Fleming's observation that a particular mould, that had drifted by chance onto an experimental Petri dish, was able to kill the bacteria growing there. It was this random event that led to the discovery of penicillin and the development of the whole field of antibiotics. Millions of lives have been saved as a result of this apparently chance discovery.

Alexander Fleming won the Nobel Prize for Physiology or Medicine in 1945, along with Ernst Chain and Howard Florey who worked out how to make penicillin in large quantities so that it could be used to treat patients. Isaac Asimov's famous statement at the top of this page flags up to us that Alexander Fleming wasn't simply some fortunate man who struck lucky. His insight wasn't a fluke. It's very unlikely that Fleming was the first scientist whose bacterial cultures had become infected with mould. His achievement came in recognising that something unusual had happened, and appreciating its significance. Knowledge and training had prepared Fleming's mind to make the most of the chance event. He saw what probably many others had seen before him but he thought what nobody else had thought.

Even if we accept the role that odd events have played in research, it would still be very comforting to think that science generally proceeds in a logical and ordered fashion. Here's one way we could imagine such progress in epigenetics …

Epigenetic modifications control cell fate – it's these processes by which liver cells, for example, stay as liver cells and don't turn into other cell types. Cancer represents a breakdown in normal control of cell fate, because liver cells stop being liver cells and become cancer cells, suggesting that epigenetic regulation has become abnormal in cancer. We should therefore aim to develop drugs that influence this epigenetic mis-regulation. Such drugs may be useful for treating or controlling cancer.

That's a neat and tidy process, and makes a lot of sense. In fact, hundreds of millions of dollars are being spent in the global pharmaceutical industry to develop epigenetic drugs for exactly this purpose. But the clear-cut thought process outlined above is not how this process of cancer drug discovery started.

There are already licensed drugs which treat cancer and which work by inhibiting epigenetic enzymes. These compounds were shown to be active against cancer cells *before* they were shown to work on epigenetic enzymes. In fact, it's the success of these compounds that has really stirred up interest in epigenetic therapies, and in the whole field of epigenetics itself – so much for a neat narrative arc.

The accidental epigeneticist

Back in the early 1970s, a young South African scientist called Peter Jones was working with a compound called 5-azacytidine. This compound was already known to have anti-cancer effects because it could stop leukaemia cells from dividing, and had some beneficial effects when tested in childhood leukaemia patients[1].

Peter Jones is now recognised as the founding father of epigenetic treatments for cancer. Tall, thin, tanned and with thick close-cropped white hair, he is an instantly recognisable presence at any conference. Like so many of the terrific scientists mentioned in this book, he has researched for decades in an ever-evolving field. He remains at the forefront of efforts to understand the impact of the epigenome on health. He is currently spearheading efforts to characterise all the epigenetic modifications present in a vast number of different cell types and diseases. These days he is able to call on technologies that allow his team to analyse millions of read-outs from highly specific and specialised equipment. Back in the early 1970s, he made his first breakthrough by being incredibly observant and thorough – a classic case of a prepared mind.

Forty years ago, nobody was quite sure how 5-azacytidine worked. It's very similar in chemical structure to base C (cytidine) from DNA and RNA. It was assumed that 5-azacytidine got added into DNA and RNA chains. Once there, it somehow disrupted normal copying of DNA, and transcription or activity of RNA. Cancer cells such as the ones found in leukaemia are extremely active. They need to synthesise lots of proteins, which means they need to transcribe a lot of mRNA. Because they divide quickly they also need to replicate their DNA very efficiently. If 5-azacytidine was interfering with one or both of these processes, it would probably hamper the growth and division of the cancer cells.

Peter Jones and his colleagues were testing the effects of 5-azacytidine on a range of cells from mammals. It's remarkably fiddly to get many types of cells to grow in the laboratory if you just take them straight out of a human or another animal. Even when you can get them to grow, they often stop dividing after a few cell divisions and die off. To get around this, Peter Jones worked with cell lines. Cell lines are derived originally from animals, including humans, but as a result of chance or experimental manipulation, they are able to grow indefinitely in culture, if given the right nutrients, temperature and environmental conditions.

Cell lines are not exactly the same as cells in the body, but they are a useful experimental system.

The type of cells that Peter Jones and his colleagues were testing are usually grown in a flat plastic flask. This looks a little like a see-through version of a hip flask for whisky or brandy, lying on its side. The mammalian cells grow on the flat inside surface of the flask. They form a single layer of cells, tightly packed side by side, but never growing on top of one another.

One morning, after the cells had been cultured with 5-azacytidine for several weeks, the researchers found that there was a strange lumpy bit in one of the culture flasks. To the naked eye, this initially looked like a mould infection. Most people would just discard the flask and make a silent promise to be a bit more careful when culturing their cells in future, to stop this happening again. But Peter Jones did something else. He looked at the lump more closely and discovered it wasn't a stray bit of mould at all. It was a big mass of cells, which had fused to form giant cells containing lots of nuclei. These were little muscle fibres, the syncytial tissue we met in the discussion of X inactivation. Sometimes the little muscle fibres would even twitch[2].

This was very odd indeed. Although the cell line had originally been derived from a mouse embryo, it never usually formed anything like a muscle cell. It tended instead to form epithelial cells – the cell type that lines the surfaces of most of our organs. Peter Jones' work showed that 5-azacytidine could change the potential of these embryonic cells, and force them to become muscle cells, instead of epithelial cells. But why would a compound that killed cancer cells, presumably by disrupting production of DNA and mRNA, have an effect like this?

Peter Jones carried on working on this when he moved from South Africa to the University of Southern California. Two years later, he and his PhD student Shirley Taylor showed that cell lines treated with 5-azacytidine didn't only form muscle. They could also form other cell types. These included fat cells (adipocytes) and cells called chondrocytes. These produce cartilage proteins, such as those that line the surfaces of joints so that the two planes can glide smoothly over each other.

These data showed that 5-azacytidine wasn't a special muscle-specifying factor. Very presciently, Professor Jones made the suggestion in his paper reporting this work that, '5-azacytidine … causes a reversion to a more pluripotent state'[3]. In other words, this compound was pushing the ball a little way back up Waddington's epigenetic landscape. The ball was then rolling back down the valleys between the hills, into a different final resting place.

But there was still no theory as to why 5-azacytidine had this unusual effect. Peter Jones himself tells a lovely self-deprecating story about the turning point in our understanding. His original appointment at the University of Southern California was in the Department of Paediatrics, but he wanted a joint appointment with the Department of Biochemistry. Part of the procedure for obtaining this joint appointment included an extra interview, which he considered quite pointless. Peter Jones described his work with 5-azacytidine in this interview and explained that no-one knew why the compound affected cell pluripotency. Robert Stellwagen, another scientist at the same university who was taking part in the interview asked, 'Have you thought of DNA methylation?'. Our candidate admitted he not only hadn't thought of it, he hadn't even heard of it[4].

Peter Jones and Shirley Taylor immediately began to focus on DNA methylation and in a very short time showed that this was indeed key to the effects of 5-azacytidine. 5-azacytidine inhibited DNA methylation. Peter Jones and Shirley Taylor created a number of related compounds and tested them for their effects in cell culture. The ones that inhibited DNA methylation also caused the changes in phenotype originally observed for 5-azacytidine. Compounds that didn't inhibit DNA methylation had no effect on phenotype[5].

The methylation cul-de-sac

Cytidine (base C) and 5-azacytidine are very similar in chemical structure. They are shown in Figure 11.1, which for simplicity only shows the most relevant parts of the structure (called cytosine and 5-azacytosine, respectively).

The top half of the diagram is very similar to Figure 4.1, showing that cytosine can be methylated by a DNA methyltransferase (DNMT1, DNMT3A or DNMT3B) to create 5-methylcytosine. In 5-azacytosine, a nitrogen atom (N) replaces the key carbon atom (C) that normally gets methylated. The DNA methyltransferases can't add a methyl group to this nitrogen atom.

Thinking back to Chapter 4, imagine a methylated region of DNA. When a cell divides, it separates the two strands of the

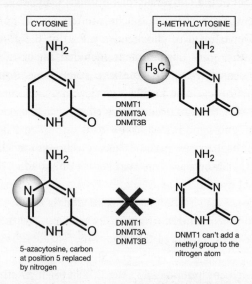

Figure 11.1   5-azacytosine can be incorporated into DNA during the DNA replication which takes place prior to cell division. 5-azacytosine takes the place of a C base, but because it contains a nitrogen atom where there is usually a carbon atom, the foreign base cannot be methylated by DNMT1 in the way that was described in Figure 4.2.

DNA double helix and copies each one. But the enzymes that copy the DNA can't themselves copy DNA methylation. As a consequence each new double helix had one methylated strand and one unmethylated one. The DNA methyltransferase called DNMT1 can recognise DNA which has only got DNA methylation on one strand and can replace it on the other strand. This restores the original DNA methylation pattern.

But if dividing cells are treated with 5-azacytidine, this abnormal cytidine base is added into the new strand of DNA as the genome gets copied. Because the abnormal base contains a nitrogen atom instead of a carbon atom, the DNMT1 enzyme can't replace the missing methyl group. If this continues as the cells keep dividing, the DNA methylation begins to get diluted out.

Something else also happens when dividing cells are treated with 5-azacytidine. We now know that when DNMT1 binds at a region where the DNA contains 5-azacytidine instead of the normal cytidine, the DNMT1 becomes stuck there[6]. This marooned enzyme is then sent to a different part of the cell and is broken down. Because of this, the total levels of DNMT1 enzyme in the cell fall[7,8]. The combination of this decrease in the amount of DNMT1, and the fact that 5-azacytidine can't be methylated, means that the amount of DNA methylation in the cell keeps dropping. We'll come back in a little while to why this drop in DNA methylation has an anti-cancer effect.

So, 5-azacytidine is an example of where an anti-cancer agent was unexpectedly shown to work epigenetically. Bizarrely, a rather similar thing happened with our second example of a compound which is now licensed to treat cancer[9].

Another happy accident

In 1971 the scientist Charlotte Friend showed that a very simple compound called DMSO (its full name is dimethyl sulfoxide) had an odd effect on the cancer cells from a mouse model of leukaemia.

When these cells were treated with DMSO, they turned red. This was because they had switched on the gene for haemoglobin, the pigment that gives red blood cells their colour[10]. Leukaemia cells normally never switch on this gene and the mechanism behind this effect of DMSO was completely unknown.

Ronald Breslow at Columbia University and Paul Marks and Richard Rifkind at Memorial Sloan-Kettering Cancer Center were intrigued by Charlotte Friend's research. Ronald Breslow began to design and create a new set of chemicals, using the structure of DMSO as his starting point, and then adding or changing bits, a little like making new combinations of Lego bricks. Paul Marks and Richard Rifkind began to test these chemicals in various cell models. Some of the compounds had a different effect from DMSO. They stopped cells from growing.

After many iterations, learning from each new and more complicated set of structures, the scientists created a molecule called SAHA (suberoylanilide hydroxamic acid). This compound was really effective at stopping growth and/or causing cell death in cancer cell lines[11]. However, it was another two years before the team were able to identify what SAHA was doing in cells. The key moment happened more than 25 years after Charlotte Friend's breakthrough publication, when Victoria Richon in Paul Marks' team, read a 1990 paper from a group at the University of Tokyo.

The Japanese group had been working on a compound called Trichostatin A or TSA. TSA was known to be able to stop cells proliferating. The Japanese group showed that treatment with TSA altered the extent to which histone proteins are decorated with the acetyl chemical group in cancer cell lines. Histone acetylation is another epigenetic modification that we first met in Chapter 4. When cells were treated with TSA, the levels of histone acetylation went up. This wasn't because the compound was activating the enzymes that put the acetyl groups on histones. It was because TSA was inhibiting the enzymes that remove acetyl groups from these chromatin proteins. These proteins are called histone deacetylases, or HDACs for short[12].

Victoria Richon compared the structure of TSA with the structure of SAHA, and the two are shown in Figure 11.2.

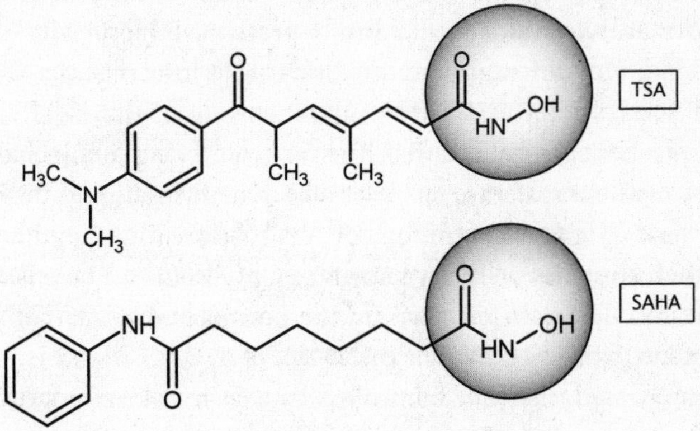

Figure 11.2 The structures of TSA and SAHA, with the areas of greatest similarity circled. C: carbon; H: hydrogen; N: nitrogen; O: oxygen. For simplicity, some carbon atoms have not been explicitly shown, but are present where there is a junction of two lines.

You don't need a chemistry degree to see that TSA and SAHA look fairly similar, especially at the right hand side of each molecule. Victoria Richon hypothesised that, just like TSA, SAHA was also an HDAC inhibitor. In 1998, she and her colleagues published a paper that showed this was indeed the case[13]. SAHA prevents HDAC enzymes from removing acetyl groups from histone proteins, and as a result, the histones carry lots of acetyl groups.

Beyond coincidence

So, 5-azacytidine and SAHA both decrease cancer cell proliferation, and both inhibit the activity of epigenetic enzymes. Although we could take this as promising support for the theory that epigenetic proteins are important in cancer, perhaps we could just be leaping to conclusions? It might just be a coincidence that both drugs affect epigenetic proteins. After all, the enzymes targeted by the two compounds are very different. 5-azacytidine inhibits the DNMT enzymes, which add methyl groups to DNA. SAHA, on the other hand, inhibits the HDAC family of enzymes, which remove acetyl groups from histone proteins. Superficially, these seem like very different processes. Maybe it's just coincidence that both 5-azacytidine and SAHA inhibit epigenetic enzymes?

Epigeneticists believe that it is far from being a coincidence. DNA methyltransferase enzymes add a methyl group to the cytidine base. High concentrations of this base are found in the long CG-rich stretches of DNA known as CpG islands. These islands are found upstream of genes, in the promoter regions that control gene expression. When the DNA of a CpG island is heavily methylated, the gene controlled by that promoter is switched off. In other words, DNA methylation is a repressive modification. DNMT activity increases DNA methylation and therefore represses gene expression. By inhibiting these enzymes with 5-azacytidine, we can drive gene expression up.

Histone proteins are also found at the promoters of genes. Histone modifications can be very complex, as we saw in Chapter 4. But histone acetylation is the most straightforward in terms of its effects on gene expression. If the histones upstream of a gene are heavily acetylated, the gene is likely to be highly expressed. If the histones are lacking acetylation, the gene is likely to be switched off. Histone *de*acetylation is a repressive change. Histone deacetylases (HDACs) remove the acetyl groups from histone proteins and will therefore repress gene expression. By inhibiting these enzymes with SAHA, we can drive gene expression up.

So there is a consistent finding. Our two unrelated compounds, which control growth of cancer cells in culture and which have now been licensed for use in human treatment, inhibit epigenetic enzymes. In doing so, they both drive up gene expression which raises the obvious question of why this is useful for treating cancer. To understand this, we need to get to grips with some cancer biology.

Cancer biology 101

Cancer is the result of abnormal and uncontrolled proliferation of cells. Normally, the cells of our body divide and proliferate at exactly the right rate. This is controlled by a complex balancing act between networks of genes in our cells. Certain genes promote cell proliferation. These are sometimes referred to as proto-oncogenes. They were represented by a plus sign in the see-saw diagram in the previous chapter. Other genes hold the cell back, preventing too much proliferation. These genes are called tumour suppressors. They were represented by a negative sign on the same diagram.

Proto-oncogenes and tumour suppressors are not intrinsically good or bad. In healthy cells, the activities of these two classes of genes balance each other. But when regulation of these networks goes wrong, cell proliferation may become mis-regulated. If a proto-oncogene becomes over-active, it may push a cell towards a cancerous state. Conversely, if a tumour suppressor gets inactivated, it will no longer act as a brake on cell division. The outcome is the same in both cases – the cell may begin to proliferate too rapidly.

But cancer isn't just a result of too much cell proliferation. If cells divide too quickly but are otherwise normal, they form structures called benign tumours. These may be unsightly and uncomfortable but unless they press on a vital organ and affect its activity, they are unlikely in themselves to be fatal. In full-blown cancer the cells don't just divide too often, they are also abnormal and can start to invade other tissues.

A mole is a benign tumour. So is a little outgrowth in the inside of the large intestine, called a polyp. Neither a mole nor a polyp is dangerous in itself. The problem is that the more of these moles or polyps you have, the greater the likelihood that one of them will go the next step, and develop an abnormality that will take it further along the path towards full-blown cancer.

This implies something rather important, that has been demonstrated in a large number of studies. Cancer is not a one-off event. Cancer is a multi-step process, where each additional step takes a cell further along the road to becoming malignant. This is true even in cases where patients inherit a very strong pre-disposition to cancer. One example is pre-menopausal breast cancer, which runs in some families. Women who inherit a mutated copy of a gene called *BRCA1* are at very high risk of early and aggressive breast cancer, which is difficult to treat effectively. But even these women aren't born with active breast cancer. It takes many years before the cancer develops, because other defects have to accumulate as well.

So, cells accumulate defects as they move increasingly close to becoming cancerous. These defects must be transmitted from mother cell to daughter cell, because otherwise they would be lost each time a cell divided. These defects must be heritable as the cancer develops. Understandably, for a very long time, the attention of the scientific community focused on identifying mutations in the genes involved in the development of cancer. They were looking for alterations in the genetic code, the fundamental blueprint. They were particularly interested in the tumour suppressor genes as these are the genes that are usually mutated in the inherited cancer syndromes.

Humans tend to have two copies of each tumour suppressor gene, as most are carried on the autosomes. As a cell becomes increasingly cancerous, both copies of key tumour suppressor genes usually get switched off (inactivated). In many cases this may be because the gene has mutated in the cancer cells. This is known as somatic mutation – it has happened in body cells at some point during normal life. These are called somatic mutations to distinguish them from genetic mutations, the ones that are transmitted from parent to child. The mutations that inactivate the two copies of a tumour suppressor may be quite variable. In some cases there may be changes in the amino acid sequence, so that the gene can't produce a functional protein any more. In other cases, there may be loss of the relevant part of the chromosome in the increasingly cancerous cells. In an individual patient, one copy of a specific tumour suppressor may carry a mutation that changes the amino acid sequence and the other may have suffered a micro-deletion.

It's abundantly clear that these events do happen, and quite frequently, but often it's been difficult to identify exactly how a tumour suppressor has mutated. In the last fifteen years, we've started to realise that there is another way that a tumour suppressor gene can become inactivated. The gene may be silenced epigenetically. If the DNA at the promoter becomes excessively methylated or the histones are covered in repressive modifications, the tumour suppressor will be switched off. The gene has been inactivated without changing the underlying blueprint.

The epigenetic frontier in cancer

Various labs have identified cancers where this has clearly happened. One of the first reports was in a type of kidney cancer called clear-cell renal carcinoma. A key step in the development of this kind of cancer is the inactivation of a specific tumour suppressor gene called *VHL*. In 1994, a group headed by the hugely influential Stephen Baylin from Johns Hopkins Medical Institution in Baltimore analysed the CpG island in front of the *VHL* gene. In 19 per cent of the clear-cell renal carcinoma samples that they analysed, the DNA of the island was hypermethylated. This switched off expression of this key tumour suppressor gene, and was almost certainly a major event in cancer progression in these individuals[14].

Promoter methylation was not restricted to the *VHL* tumour suppressor and renal cancer. Professor Baylin and colleagues subsequently analysed the *BRCA1* tumour suppressor gene in breast cancer. They analysed cases where there was no family history of this disease, and the cancer wasn't caused by the mutations in *BRCA1* that we discussed a few paragraphs ago. In 13 per cent of these sporadic cases of breast cancer, the *BRCA1* CpG island was hypermethylated[15]. Broader abnormal patterns of DNA methylation in cancer were reported by Jean-Pierre Issa from the MD Anderson Cancer Center in Houston, in collaboration with Stephen Baylin. Their collaborative work showed that over 20 per cent of colon cancers had high levels of promoter DNA methylation, at many different genes simultaneously[16].

Follow-on work showed that it's not just DNA methylation that changes in cancer. There is also direct evidence for histone modifications leading to repression of tumour suppressor genes. For example, the histones associated with a tumour suppressor gene called *ARHI* had low levels of acetylation in breast cancer[17]. A similar relationship exists for the *PER1* tumour suppressor in a form of lung cancer called non-small cell[18]. In both cases, there was a relationship between the levels of histone acetylation and the expression of the tumour suppressor – the lower the levels of acetylation, the lower the expression of the gene. Because these genes are both tumour suppressors, their decreased expression would mean that the cell would find it harder to put the brakes on proliferation.

This realisation – that tumour suppressor genes are often silenced by epigenetic mechanisms – has led to considerable excitement in the field, because this potentially creates a new way of treating cancer. If you can turn one or more tumour suppressor genes back on in cancer cells, there is a fighting chance of reining in the crazy proliferation rate of those cells. The runaway train may not run away quite so fast down the track.

When scientists thought that tumour suppressors were inactivated by mutations or deletions, we didn't have many options for turning these genes back on. There are trials in progress to test if gene therapy can be used to achieve this. There may be circumstances where gene therapy will prove effective, but this is by no means certain. Gene therapy has struggled to deliver on the initial hopes for this technology, in all sorts of diseases. It can be very difficult to get the genes delivered into the right cells, and to get them to switch on when they are there. Even when we're able to do this, we often find that the body gets rid of these extra genes, so any initial benefit is lost. There have also been relatively rare cases where the gene therapy itself has led to cancer, because it has had unexpected effects which have led to increased cell proliferation. The scientific community hasn't given up hope for gene therapy and for some conditions it may yet prove to be the right approach[19]. But for diseases like cancer, where we would need to treat a lot of people, it's expensive and difficult.

That's why there is so much excitement about the development of epigenetic drugs to treat cancer. By definition, epigenetic changes do not alter the underlying DNA code. As we have seen, there are patients where one copy of a tumour suppressor has been silenced by the action of epigenetic enzymes. In these patients the code for the normal tumour suppressor protein has not been corrupted by mutation. So, for them there is the possibility that treatment with appropriate epigenetic drugs can reverse the abnormal pattern of DNA methylation or histone acetylation. If we can achieve this, the normal tumour suppressor gene will be switched back on, and this will help bring the cancer cells back under control.

Two drugs that inhibit the DNMT1 enzyme have been licensed for clinical use in cancer patients by the Food and Drug Administration (FDA) in the USA. These are 5-azacytidine (tradename *Vidaza*) and the closely related 2-aza-5′-deoxycytidine (tradename *Dacogen*). Two HDAC inhibitors have also been licensed. These are SAHA (tradename *Zolinza*), which we met earlier, and a molecule called romidepsin (tradename Istodax), which has a very different chemical structure from SAHA, but which also inhibits HDAC enzymes.

Following on from his successes in unravelling the molecular roles of 5-azacytidine, Peter Jones, along with Stephen Baylin and Jean-Pierre Issa, has played a hugely influential role in the last 30 years in moving this compound from the laboratory, all the way through clinical trials and finally to the licensed product. Victoria Richon played a major role in championing SAHA all the way through the same process.

The successful licensing of these four compounds against two different types of enzymes has given a major boost to the whole field of epigenetic therapies. But they have not proved to be universal wonder drugs, the silver bullets to treat all cancers.

Stop looking for miracles

That hasn't been a surprise to anyone working in the fields of cancer research and treatment. There sometimes seems to be an obsessive determination on the part of certain journalists in the popular press to write about *the* cure for cancer. Generally speaking, scientists try to avoid being too dogmatic, but if there's one thing most of them are agreed on, it's that there will never be one single cure for cancer.

That's because there isn't one form of cancer. There are probably over a hundred different diseases with this name. Even if we take just one example – say breast cancer – we find that there are different types of this particular strain of cancer. Some grow in response to the female hormone called oestrogen. Some respond most strongly to a protein called epidermal growth factor. The *BRCA1* gene is inactivated or mutated in some breast cancer cases, but not in others. Some breast cancers don't respond to any of the known cancer growth factors but to some other signals which we may not even be able to identify yet.

Because cancer is a multi-step process, two patients whose cancers appear very similar may be ill because of very different molecular processes. Their cancers may have rather different combinations of mutations, epigenetic modifications and other factors driving the growth and aggressiveness of the tumour. This means that different patients are likely to require different types and combinations of anti-cancer drugs.

Even allowing for this, however, the results from clinical trials with DNMT and HDAC inhibitors have been surprising. Neither of them has yet been shown to work well in solid tumours such as cancers of the breast, colon or prostate. Instead, they are most effective against cancers that have developed from cells that give rise to the circulating white blood cells that are part of our defences against pathogens. These are referred to as haematological tumours. It's not clear why the current epigenetic drugs don't seem to be effective against solid tumours. It might be that there are different molecular mechanisms at work in these, compared with haematological cancers. Alternatively, it could be that the drugs can't get into solid tumours at high enough concentrations to affect most of the cancer cells.

Even within haematological tumours, there are differences between the DNMT and HDAC inhibitor drugs. Both DNMT inhibitors have been licensed for use in a condition called myelodysplastic syndrome[20,21]. This is a disorder of the bone marrow.

Both HDAC inhibitors have been licensed for a different kind of haematological tumour, called cutaneous T cell lymphoma[22]. In this disease, the skin becomes infiltrated with proliferating immunological cells called T cells, creating visible plaques and large lesions.

Not every patient with myelodysplastic syndrome or cutaneous T cell lymphoma gains a clinical benefit from taking these drugs. Even amongst the patients who do respond, none of these drugs really seem to cure the condition. If the patients stop taking the drugs, the cancer regains its hold. The DNMT1 inhibitors and the HDAC inhibitors seem to rein in the cancer cell growth, retarding and repressing it. They control rather than cure.

However, this often represents a significant improvement for the patients, bringing prolonged life expectancy and/or improved quality of life. For example, many patients with cutaneous T cell lymphoma suffer significant pain and distress because their lesions are constantly and excruciatingly itchy. The HDAC inhibitors are often very effective at calming this aspect of the cancer, even in patients whose survival times aren't improved by these drugs.

Generally speaking, it's often very difficult to know which patients will benefit from a specific new anti-cancer drug. This is one of the biggest problems facing the companies working on new epigenetic therapies for the treatment of cancer. Even now, several years after the first licences were granted by the FDA for 5-azacytidine and SAHA, we still don't know why they work so much better in myelodysplastic syndrome and cutaneous T cell lymphoma than in other cancers. It just so happened that in the early clinical trials in humans, patients who had these conditions responded more strongly than patients with other types of cancers. Once the clinicians running the trials noticed this, later trials were designed that focused around these patient groups.

This may not sound like a major difficulty. It might seem straightforward for companies to develop drugs and then test them in all sorts of cancers and with all sorts of combinations of other cancer drugs, to work out how to use them best.

The problem with this is the expense. If we check out the website of the National Cancer Institute, we can look for the number of trials that are in progress for a specific drug. In February 2011, there were 88 trials to test SAHA[23]. It's difficult to get definitive costs for how much clinical trials cost, but based on data from 2007, a value of $20,000 per patient is probably a conservative estimate[24]. Assuming each trial contains twenty patients, this would mean that the costs just for testing SAHA in the trials at the National Cancer Institute are over $35,000,000. And this is almost certainly an under-estimate of the overall cost.

The researchers at Columbia University and Memorial Sloan-Kettering who first developed SAHA patented it. They then set up a company called Aton Pharma to develop SAHA as a drug. In 2004, after promising early results in cutaneous T cell lymphoma, Aton Pharma was bought by the giant pharmaceutical company Merck for over $120 million dollars. Aton Pharma had almost certainly spent millions of dollars to get SAHA to this stage. Drug discovery and development is an expensive business. The two companies that marketed the DNMT1 inhibitors have been bought relatively recently by larger pharmaceutical companies, in deals that totalled about $3 billion each[25]. If a company has paid a huge amount of money to develop or buy in a new drug, it would much prefer not to carry on spending like a drunken sailor when it comes to clinical trials.

Naturally, it would be a big improvement if we could run clinical trials with a much better idea of which patients will benefit, rather than having to take pot luck. Unfortunately, most researchers agree that many of the animal models used to test cancer drugs are very limited in their capacity to predict the most susceptible human cancer. To be fair, this isn't just true of cancer drugs targeted at epigenetic enzymes, it's also true of pretty much all oncology drug discovery.

In an attempt to get around this problem, researchers in both academia and industry are now looking for the next generation of epigenetic targets in oncology. DNMT1 is a relatively broad-acting enzyme. DNA methylation is rather all or nothing – a CpG is methylated or it isn't. HDACs tend to be pretty non-discriminating too. If they can get access to an acetylated lysine on a histone tail, they'll take that acetyl group off. There are a lot of lysines on a histone tail – there are are seven on histone H3, just for starters. There are at least ten different HDAC enzymes that SAHA can inhibit. It's quite likely that each of these ten can deacetylate any of the seven lysines on the H3 tail. This is hardly what we would call fine-tuning.

No easy wins

This is why the field is now moving in the direction of assessing different epigenetic enzymes, which are much more limited in their actions, to see which are important players in different cancers. The rationale is that it will be easier to understand the cellular biology of enzymes with quite limited actions, and this will make it easier to determine which patients are likely to respond best to which drugs.

The first problem in doing this is rather a daunting one. Which proteins should we investigate? There are probably at least a hundred enzymes that add or remove histone modifications (writers and erasers of the epigenetic code). There are probably as many proteins that read the epigenetic code. To make matters worse, many of these writers, erasers and readers interact with each other. How can we begin to identify the most promising candidates for new drug discovery programmes?

We don't have any useful compounds like 5-azacytidine and SAHA to guide us, so we have to rely on our relatively incomplete knowledge in cancer and in epigenetics. One area that is proving useful is considering how histone and DNA modifications work in tandem.

The most heavily repressed areas of the genome have high levels of DNA methylation and are extremely compacted. The DNA has become very tightly wound up, and is exceptionally inaccessible to enzymes that transcribe genes. But it's the question of how these regions become so heavily repressed that is really important. The model is shown in Figure 11.3.

In this model, there is a vicious cycle of events that results in the generation of a more and more repressed state. One of the predictions from this model is that repressive histone modifications attract DNA methyltransferases, which deposit DNA methylation near those histones. This methylation in turn attracts more repressive histone modifying enzymes, creating a self-perpetuating cycle that leads to an increasingly hostile region for gene expression.

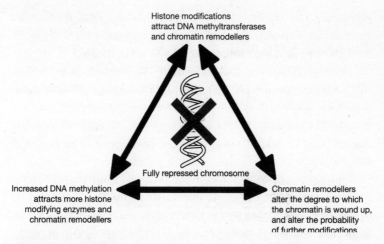

Histone modifications
attract DNA methyltransferases
and chromatin remodellers

Fully repressed chromosome

Increased DNA methylation
attracts more histone
modifying enzymes and
chromatin remodellers

Chromatin remodellers
alter the degree to which
the chromatin is wound up,
and alter the probability
of further modifications

Figure 11.3 Schematic to illustrate how different types of epigenetic modifications act together to create an increasingly repressed and tightly condensed chromosome region, making it very difficult for the cell to express genes from this region.

Experimental data suggest that in many cases this model seems to be right. Repressive histone modifications can act as the bait to attract DNA methylation to the promoter of a tumour suppressor gene. A key example of this is an epigenetic enzyme we met in the previous chapter, called EZH2. The EZH2 protein adds methyl groups to the lysine amino acid at position 27 on histone H3. This amino acid is known as H3K27. K is the single letter code for lysine (L is the code for a different amino acid called leucine).

This H3K27 methylation itself tends to switch off gene expression. However, in at least some mammalian cell types, this histone methylation recruits DNA methyltransferases to the same region of chromatin[26,27]. The DNA methyltransferases include DNMT3A and DNMT3B. This is important because DNMT3A and DNMT3B can carry out the process known as de novo DNA methylation. That is, they can methylate virgin DNA, and create completely new regions of highly repressed chromatin. As a result, the cell can convert a relatively unstable repressive mark (H3K27 methylation) into the more stable DNA methylation.

Other enzymes are also important. An enzyme called LSD1 takes methyl groups off histones – it's an eraser of epigenetic modifications[28]. It does this particularly strongly at position 4 on histone H3 (H3K4). H3K4 is the opposite of H3K27, because when H3K4 is clear of methyl groups, genes tend to be switched off.

Unmethylated H3K4 can bind proteins, and one of these is called DNMT3L. Perhaps not surprisingly, this is related to DNMT3A and DNMT3B. DNMT3L doesn't methylate DNA itself, but it attracts DNMT3A and DNMT3B to the unmethylated H3K4. This provides another way to target stable DNA methylation to virgin territory[29].

In all likelihood, many histones positioned at the promoters of tumour suppressor genes carry both of these repressive histone marks – methylation of H3K27 and non-methylation of H3K4 – and these act together to target the DNA methyltransferases even more strongly.

Both EZH2 and LSD1 are up-regulated in certain cancer types, and their expression correlates with the aggressiveness of the cancer and with poor patient survival[30,31]. Basically, the more active these enzymes, the worse the prognosis for the patient.

So, histone modifications and DNA methylation pathways interact. This may explain, at least in part, one of the mysteries of existing epigenetic therapies. Why are compounds like 5-azacytidine and SAHA only *controllers* of cancer cells, rather than complete destroyers?

In our model, treatment with 5-azacytidine will drive down the DNA methylation for as long as the patients take the drug. Unfortunately, many cancer drugs have serious side-effects and the DNMT inhibitors are no exception. The side effects may eventually become such a problem that the patient has to stop taking the drug. However, the patient's cancer cells probably still have histone modifications at the tumour suppressor genes. Once the patient stops taking 5-azacytidine, these histone modifications almost certainly start to attract the DNMT enzymes all over again, re-initiating stable repression of gene expression.

Some researchers are carrying out clinical trials using 5-azacytidine and SAHA together to try to interfere with this cycle, by disrupting both the DNA and histone components of epigenetic silencing. It's not clear yet if these will be successful. If they aren't, it might suggest that it's not low levels of histone acetylation which are most important for re-establishing the DNA methylation. It might be that specific histone modifications, of the types just described, are more important. But we don't yet have drugs to inhibit any of the other epigenetic enzymes, so we're stuck with Hobson's choice at the moment, that is, no choice at all.

In the future, we may not need to use DNMT inhibitors at all. The link between DNA methylation and histone modifications in cancer isn't absolute. If a CpG island is methylated, the downstream gene is repressed. But there are tumour suppressor genes that are downstream of unmethylated CpG islands and tumour suppressor genes that don't have a CpG island at all. These genes may still be repressed, but solely thanks to histone modifications[32]. This has been shown by Jean-Pierre Issa at the MD Anderson Cancer Center in Houston, who has been so instrumental in the implementation of epigenetic therapies in the clinic. In these instances, if we can find the right epigenetic enzymes to target with inhibitors, we may be able to drive re-expression of the tumour suppressors without needing to worry about DNA methylation.

An uneasy truce

Is there something special about the tumour suppressor genes that get silenced using epigenetic modifications? There are two contrasting theories about this. The first is that there's nothing special about these genes and the process is completely random. In this model, every once in a while a random tumour suppressor gets abnormally modified epigenetically. If this changes the expression of the gene, it may mean that cells with that epigenetic modification grow a bit faster or a bit better than their neighbours. This gives the cells a growth advantage and they keep outgrowing the cells around them, gradually accumulating more epigenetic and genetic changes that make them ever more cancerous.

The other theory is that the tumour suppressors that become repressed epigenetically are somehow targeted in this process. It's not just random bad luck, these genes are actually at a higher than average risk of epigenetic silencing.

In recent years, as we've had the technology to profile the epigenetic modifications in more and more cell types, and at higher and higher resolutions, the field has shifted in favour of the second model. There are a set of genes that seem to be rather prone to getting themselves switched off epigenetically.

At first this seems incredibly counter-intuitive. Why on earth would billions of years of evolution leave us with cellular systems that render us prone to cancerous changes? Well, we have to put this in context. Most evolutionary pressures are connected with the drive to leave behind as many offspring as possible. For a human to reach reproductive age, it's incredibly important that early development occur as efficiently as possible. After all, you can't reproduce if you never make it past the embryo stage. Once we've reached reproductive age and had the opportunity to reproduce, there is little to be gained in evolutionary terms in us staying alive for several decades afterwards.

Evolution has favoured cellular mechanisms that promote effective early growth and development, including the production of multiple different tissue types. Many of these tissue types contain reservoirs of stem cells which are specific to that tissue. Our bodies need these for tissue growth as we mature, and for tissue regeneration following injury. The fates and identities of these tissue-specific stem cells are controlled by the precise patterns of epigenetic modifications. By using epigenetic modifications to control gene expression, the cells keep some flexibility. They have the potential to change into more specialised cells, for example. Perhaps even more importantly when considering cancer, the epigenetic modifications also allow cells to divide to form more stem cells. This is why we don't run out of skin cells, or bone marrow cells, even if we live to be a hundred years old.

This requirement for gene expression patterns that aren't completely set in stone is probably why epigenetic repression of tumour suppressor genes is not a random process. We can't have things two ways. Regulatory systems that allow cells to be flexible are inevitably also systems that allow cells to go wrong. In evolutionary terms, it's the price we have to pay for our Goldilocks scenario. Our epigenetic pathways make sure some of our cells aren't completely pluripotent or completely differentiated. Instead, they are just right, hovering somewhere near the top of Waddington's epigenetic landscape, but ready to roll down at any time.

Peter Laird, who like Peter Jones is based at the University of Southern California, has shown the knock-on effects of this system in cancer cells. His team analysed patterns of DNA methylation in cancer cells, especially focusing on the promoters of tumour suppressor genes. Tumour suppressor genes whose histones are methylated by the EZH2 complex in ES cells were twelve times as likely to have abnormally high levels of DNA methylation as those genes that aren't targeted by EZH2. Peter Laird described this effect very elegantly, saying, 'reversible gene repression is replaced by permanent silencing, locking the cell into a perpetual state of self-renewal and thereby predisposing to subsequent malignant transformation [sic].'[33] This is consistent with the idea that there is a stem cell aspect to cancer. If cells are locked into a stem cell-like state, where they can't differentiate into cells at the bottom of the epigenetic landscape, they will be very dangerous because they will always be able to keep on dividing to form yet more cells like themselves.

Jean-Pierre Issa has described the genes that are epigenetically silenced in colon cancer as the gatekeepers. They are frequently genes whose normal role is to move the cells away from self-renewal, and into fully differentiated cell types. Inactivation of these genes in cancer locks the cells in a self-renewing stem cell-like state. This creates a population of cells that are able to keep dividing, keep accumulating further epigenetic changes and mutations, and keep inching towards a full-blown cancer state[34].

When we visualise the cells in Waddington's landscape, it's quite difficult to visualise the ones that linger somewhere near the top. That's because instinctively we know that that's a really unstable place to be. A ball that has started rolling down a slope is always likely to keep going, unless something can hold it back. And even if such a ball does come to a halt, there's always the chance it will start moving again, rolling on down that hill.

What holds cells in this teetering position? In 2006, a group headed by Eric Lander at the Broad Institute in Boston, found at least part of the answer. A key set of genes in ES cells, the pluripotent cells we have come to know so well, were found to have a really strange histone modification pattern. These were genes that were very important for controlling if an ES cell stayed pluripotent, or differentiated. Histone H3K4 was methylated at these genes, which normally is associated with switching on gene expression. H3K27 was also methylated. This is normally associated with switching off gene expression. So, which modification would turn out to be stronger? Would the genes be switched on or off?

The answer turned out to be both. Or neither, depending on which way we look at it. These genes were in a state called 'poised'. Given the slightest encouragement – a change in culture conditions that pushed cells towards differentiation for example – one or other of these methylations was lost. The gene was fully switched on, or strongly repressed, depending on the epigenetic modification[35].

This is really important in cancer. Stephen Baylin is the third person, along with Peter Jones and Jean-Pierre Issa, who has done so much to make epigenetic therapies a reality. He has shown that these poised histone modifications are found in early cancer stem cells and are really significant for setting the DNA methylation patterns in cancer cells[36].

Of course, other events must also be taking place. Many people do not develop cancer, no matter what age they live to. Something must happen in the people who do develop cancer, which results in the normal stem cell pattern getting subverted and hardened so that the cells are locked into their aggressively and abnormally proliferative state. We know that environment can have a substantial impact on cancer risk (just think of the hugely increased risk of lung cancer in smokers) but we're not clear on how or if the environment intersects with these epigenetic processes.

There may also be an aspect of pure bad luck in who develops cancer. We probably all have random fluctuations in the levels, activity or localisation of proteins that target, write, interpret and erase our epigenetic codes. And there are the non-coding RNAs too.

The 3′ UTRs of both *DNMT3A* and *DNMT3B* mRNA contain binding sites for a family of miRNAs called miR-29. Normally, these miRNAs will bind to the *DNMT3A* and *DNMT3B* mRNA molecules and down-regulate them. In lung cancer, the levels of these miRNAs drop and as a consequence *DNMT3A* and *DNMT3B* mRNA and subsequently protein expression is elevated. This is likely to increase the amount of de novo methylation of susceptible tumour suppressor promoters[37].

It is likely that there will also be feedback loops between miRNAs and the epigenetic enzymes they control, if one component of the pathway becomes mis-regulated. This will reinforce abnormal control mechanisms in the cell, leading to yet another vicious cycle, and is shown in Figure 11.4. In this example, a

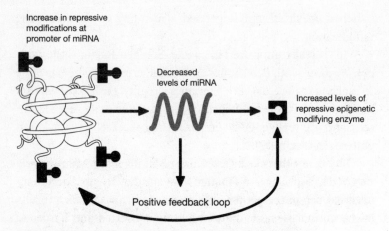

Increase in repressive
modifications at
promoter of miRNA

Decreased
levels of miRNA

Increased levels of
repressive epigenetic
modifying enzyme

Positive feedback loop

Figure 11.4   A positive feedback loop which constantly drives down
expression of a miRNA which would normally control expression of an
epigenetic enzyme that creates a repressed chromatin state.

miRNA regulates a specific epigenetic enzyme, which itself modifies the promoter of the miRNA. In this case, the epigenetic
enzyme creates a repressive modification.

There is still much that we need to understand if we are to
develop the next generation of epigenetic drugs to treat cancer
patients. We need to know which drugs will work best in which
diseases and which patients will benefit the most. We want to be
able to work this out in advance, so that we don't have to hope
for the best when running large numbers of clinical trials. At least
5-azacytidine and SAHA have given us the comfort of knowing
that epigenetic therapy is possible in cancer, even if improvements
are needed.

As we shall see in the next chapter, epigenetic problems are
not restricted to cancer. But sadly we are even further away from
knowing how to use epigenetic therapies in one of the greatest
unmet clinical needs in the western world – psychiatric disease.

Chapter 12

All in the Mind

The mind is its own place, and in itself can make a
Heaven of Hell, a Hell of Heaven.
John Milton

One of the most noticeable publishing trends of the last ten years has been the rise and rise of the 'misery memoir'. In this genre, the authors recount the tough times of their childhood and how they have risen above them to be successful and fulfilled individuals. The genre can be sub-divided into two categories. The first is the poor-but-happy tale, the 'we had nothing but we had love' story. The second, which may or may not also include poverty, tends to be much more disturbing. It focuses on harrowing tales of childhood neglect and childhood abuse, and some of these memoirs have been hugely successful. *A Child Called It* by Dave Pelzer, possibly the most famous of this category of books, spent over six years on the *New York Times* bestsellers list.

A substantial amount of their appeal seems to lie in the triumph-over-adversity aspects of these memoirs. Readers seem to take heart from the stories of individuals who, despite a terrible start in life, finally grow up to be happy, well-balanced adults. We applaud those who become winners 'against the odds'.

This tells us something quite revealing. It shows that, as a society, we believe that early childhood events are extremely important in influencing adult life. It also shows that we believe that it is very difficult to get over the effects of early trauma. As a readership, we possibly value these successful survivors because of what we perceive as their relative rarity.

In many ways, we are correct in our assumptions as it is true that dreadful early childhood experiences really can have a dramatic impact on adult life. There are all sorts of ways in which this has been measured and the precise figures may vary from study to study. Despite this, certain clear trends have emerged. Childhood abuse and neglect result in adults with a three times greater risk of suicide than the general population. Abused children are at least 50 per cent more likely than the general population to suffer from serious depression as adults, and will find it much harder to recover from this illness. Adults who were subjected to childhood abuse and neglect are also at significantly higher risk of a range of other conditions including schizophrenia, eating disorders, personality disorders, bipolar disease and generalised anxiety. They are also more likely to abuse drugs or alcohol[1].

An abusive or neglectful environment when young is clearly a major risk factor for the development of later neuropsychiatric disorders. We are so aware of this as a society that sometimes we almost forget to question why this should be the case. It just seems self-evident. But it's not. Why should events that lasted for two years, for example, still have adverse consequences for that individual several decades later?

One explanation that is often given is that the children are 'psychologically damaged' by their early experiences. Whilst true, this isn't that helpful a statement. The reason why it's not helpful is that the phrase 'psychologically damaged' isn't really an explanation at all – it's a description. It sounds quite convincing but on certain levels it doesn't really tell us anything.

Any scientist addressing this problem will want to take this description and probe it at another level. What are the molecular events that underlie this psychological damage? What happens in the brains of the abused or neglected children, that leaves them so prone to mental health problems as adults?

There is sometimes resistance to this approach from other disciplines, which work within different conceptual frameworks.

This seems rather puzzling. If we don't accept there is a molecular basis to a biological effect, what are we left with? A religious person may prefer to invoke the soul, just as a Freudian therapist may invoke the psyche. Both of these refer to a theoretical construct that has no defined physical basis. Moving into such a model system, where it is impossible to develop the testable hypotheses that are the cornerstone of all scientific enquiry, is deeply unattractive to most scientists. We prefer to probe for a mechanism that has a physical foundation, rather than defaulting to a scenario in which there is something which is assumed, somehow, to be a part of us, without having any physical existence.

This can generate a cultural clash, but it's one that's based on a misunderstanding. A scientist will expect that observable events have a physical basis. For the topic of this chapter, our proposed hypothesis is that terrible early childhood experiences change certain physical aspects of the brain during a key developmental period. This in turn affects the likelihood of mental health problems in adult life. This is a mechanistic explanation. It's lacking in details, admittedly, but we'll fill in some of these in this chapter. Mechanistic explanations often sit uncomfortably in our society, because they sound too deterministic. Mechanistic explanations are misinterpreted and taken to imply that humans are essentially robots, wired and programmed to respond in certain ways to certain stimuli.

But this doesn't have to be the case. If a system has enough flexibility, then one stimulus doesn't always have to result in the same outcome. Not every abused or neglected child develops into a vulnerable, unwell adult. A phenomenon can have a mechanistic basis, without being deterministic.

The human brain possesses sufficient flexibility to generate different adult outcomes in response to similar childhood experiences. Our brains contain one hundred billion nerve cells (neurons). Each neuron makes links with ten thousand other neurons to form an incredible three dimensional grid. This grid therefore contains a thousand trillion connections – that's 1,000,000,000,000,000 (a quadrillion). It's hard to imagine this, so let's visualise each connection as a disc that's 1mm thick. Stack up the quadrillion discs on top of each other and they will reach to the sun (which is ninety-three million miles from the earth) and back, three times over.

That's a lot of connections, so it's perfectly possible to imagine that our brains have a lot of flexibility. But the connections are not random. There are networks of cells within the giant grid which are more likely to link to each other than to anywhere else. It's this combination of huge flexibility, but constrained within certain groupings, that is compatible with a system that is mechanistic but not entirely deterministic.

The child is (epigenetically) father to the man

The reason scientists have hypothesised that the adult sequelae of early childhood abuse may have an epigenetic component is that we're dealing with scenarios where a triggering event continues to have consequences long after the trigger itself has disappeared. The long-term consequences of childhood trauma are very reminiscent of many of the effects that are mediated by epigenetic systems. We have seen some examples of this already. Differentiated cells remember what cell type they are, even after the signal that told them to become kidney cells or skin cells has long since vanished. Audrey Hepburn suffered from ill-health her whole life because of the malnutrition she suffered as a teenager during the Dutch Hunger Winter. Imprinted genes get switched off at certain stages in development, and stay off throughout the rest of life. Indeed, epigenetic modifications are the only known mechanism for maintaining cells in a particular state for exceptionally long periods of time.

The hypothesis that epigeneticists are testing is that early childhood trauma causes an alteration in gene expression in the brain, which is generated or maintained (or both) by epigenetic mechanisms. These epigenetically mediated abnormalities in gene expression predispose adults to increased risk of mental illnesses.

In recent years, scientists have begun to generate data suggesting that this is more than just an appealing hypothesis. Epigenetic proteins play an important role in programming the effects of early trauma. Not only that, they also are involved in adult depression, drug addiction and 'normal' memory.

The focus of a lot of research in this field has been a hormone called cortisol. This is produced from the adrenal glands which sit on top of the kidneys. Cortisol is produced in response to stress. The more stressed we are, the more cortisol we produce. The average level of cortisol production tends to be raised in adults who had traumatic childhoods, even if the individuals are healthy at the time of measurement[2,3]. What this shows is that adults who were abused or neglected as children have higher background stress levels than their contemporaries. Their systems are chronically stressed. The development of mental illness is, in many cases, probably a little like the development of cancer. A lot of things need to go wrong at the molecular level before a person becomes clinically ill. The chronic stress levels in the abuse survivors push them closer to that threshold. This increases their vulnerability to disease.

How does this over-expression of cortisol happen? It's a consequence of events that happen far from the kidneys, in our brains. There is a whole signalling cascade involved here. Chemicals produced in one region of the brain act on other areas. These areas in turn produce other chemicals in response and the process continues. Eventually a chemical leaves the brain and signals to the adrenal glands and cortisol is produced. During an abusive childhood, this signalling cascade is very active. In many abuse survivors, this system keeps signalling as if the person is still trapped in the abusive situation. It's as if the thermostat on a central heating system has malfunctioned, and the boiler and radiators continue to pump out heat in August, based on the weather from the previous February.

The process starts in a region of the brain called the hippocampus, which gets its name from the ancient Greek term for seahorse, it being shaped a little like this creature. The hippocampus acts as a master switch in controlling how much the cortisol system becomes activated. This is shown in Figure 12.1. In this figure, a plus symbol indicates that one event acts to stimulate the next link in the chain. A minus symbol shows the opposite effect, where one event decreases the level of activity of the next event in the chain.

Because of changes in the activities of the hippocampus in response to stress, the hypothalamus produces and releases two hormones, called corticotrophin-releasing hormone and arginine

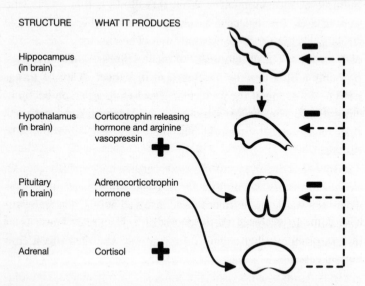

STRUCTURE   WHAT IT PRODUCES

Hippocampus
(in brain)

Hypothalamus   Corticotrophin releasing
(in brain)   hormone and arginine
                   vasopressin

Pituitary   Adrenocorticotrophin
(in brain)   hormone

Adrenal   Cortisol

Figure 12.1   Signalling events in response to stress set up a cascade of events in selected regions of the brain that ultimately result in release of the stress hormone cortisol from the adrenal glands. Under normal circumstances, this system is controlled by a set of negative feedback loops that act to dampen down and limit the activation of the stress response pathways.

vasopressin. These two hormones stimulate the pituitary, which responds by releasing a substance called adrenocorticotrophin hormone which gets into the bloodstream. When the cells of the adrenal gland take up this hormone, they release cortisol.

There's a clever mechanism built in to this system. Cortisol circulates around the body in the bloodstream, and some of it goes back into the brain. The three brain structures shown in our diagram all carry receptors that recognise cortisol. When cortisol binds to these receptors, it creates a signal that tells these structures to calm down. It's particularly important for this to happen at the hippocampus, as this structure can send out signals to dampen down all the others involved in this signalling. This is a classic negative feedback loop. Production of cortisol feeds back on various tissues, and the final effect is that the production of cortisol declines. This stops us from being constantly over-stressed.

But we know that adults who suffered traumatic childhoods *are* actually over-stressed. They produce too much cortisol, all the time. Something must be going wrong in this feedback loop. There are a few studies in humans that show that this is the case. These studies examined the levels of corticotrophin-releasing hormone in the fluid bathing the brain and spinal cord. As predicted, the levels of corticotrophin-releasing hormone were higher in individuals who had suffered childhood abuse than in individuals who hadn't. This was true even when the individuals were healthy at the time of the experiments[4,5]. Because it's so hard to investigate this fully in humans, a lot of the breakthroughs in this field have come from using animal models of certain conditions and then correlating them where possible with what we know from human cases.

Relaxed rats and mellow mice

A useful model has been based around the mothering skills of rats. In the first week of their lives, rat babies love being licked and groomed by their mothers. Some mothers are naturally very good at this, others not so much so. If a mother is good at it, she's good at it in all her pregnancies. Similarly, if she's a bit lackadaisical at the licking and grooming, this is true for every litter she has.

If we test the offspring of these different mothers when the pups are older and independent, an interesting effect emerges. When we challenge these now adult rats with a mildly stressful situation, the ones that were licked and groomed the most stay fairly calm. The ones that were relatively deprived of 'mother love' react very strongly to even mild stress. Essentially, the rats that had been licked and groomed the most as babies were the most chilled out as adults.

The researchers carried out experiments where newborn rats were transferred from 'good' mothers to 'bad' and vice versa. These experiments showed that the final responses of the adults were completely due to the love and affection they received in the first week of life. Babies born to mothers who were lacklustre lickers and groomers grew up nicely chilled out if they were fostered by mothers who were good at this.

The low stress levels of the adult rats that had been thoroughly nurtured as babies were shown by measuring their behaviour when they were challenged by mild stimuli. They were also monitored hormonally, and the effects were as we would expect. The chilled-out rats had lower levels of corticotrophin-releasing hormone in their hypothalamus and lower levels of adrenocorticotrophin hormone in their blood. Their levels of cortisol were also low, compared with the less nurtured animals.

The key molecular factor in dampening down the stress responses in the well-nurtured rats was the expression of the cortisol receptor in the hippocampus. In these rats, the receptor was highly expressed. As a result, the cells of the hippocampus were very efficient at catching even low amounts of cortisol, and using this as the trigger to subdue the downstream hormonal pathway, through the negative feedback loop.

This showed that levels of the cortisol receptor stayed high in the hippocampus, many months after the all-important licking and grooming of the baby rats. Essentially, events that only happened for seven days immediately after birth had an effect that lasted for pretty much all of a rat's life.

The reason the effect was so long-lasting is that the initial stimulus – being licked and groomed by the mother – set off a chain of events that led to epigenetic changes to the cortisol receptor gene. These changes occurred very early in development when the brain was at its most 'plastic'. By plastic, we mean that this is the time when it's easiest to modify the gene expression patterns and cellular activities. As the animals get older, these patterns stay set in place. That's why the first week in rats is so critical.

The changes that take place are shown in Figure 12.2. When a baby rat is licked and groomed a lot, it produces serotonin, one of the feel-good chemicals in mammalian brains. This stimulates expression of epigenetic enzymes in the hippocampus, which ultimately results in decreased DNA methylation of the cortisol receptor gene. Low levels of DNA methylation are associated with high levels of gene expression. Consequently, the cortisol receptor is expressed at high levels in the hippocampus, and can keep the rats relatively relaxed[6].

This is a very interesting model to explain how early life events can influence long-term behaviour. But it seems unlikely that just one epigenetic alteration – even one as significant as DNA methylation levels at a very important gene in a critical brain region – could be the only answer. Five years after the work described above, another paper was published by a different group. This also showed the importance of epigenetic changes but in a different gene.

The later group used a mouse model of early-life stress. In this model, baby mice were taken away from their mothers for three hours a day, for the first ten days of their lives. Just like the baby rats that hadn't been licked or groomed much, these babies

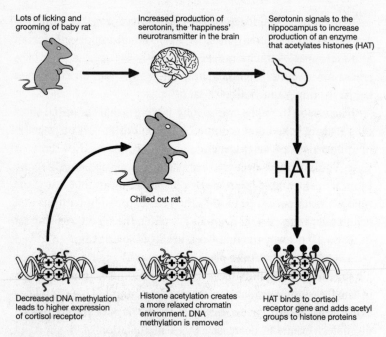

Lots of licking and grooming of baby rat

Increased production of serotonin, the 'happiness' neurotransmitter in the brain

Serotonin signals to the hippocampus to increase production of an enzyme that acetylates histones (HAT)

HAT

Chilled out rat

Decreased DNA methylation leads to higher expression of cortisol receptor

Histone acetylation creates a more relaxed chromatin environment. DNA methylation is removed

HAT binds to cortisol receptor gene and adds acetyl groups to histone proteins

Figure 12.2 Strong nurturing of baby rats sets up a cascade of molecular events that result in increased expression of the cortisol receptor in the brain. This increased expression makes the brain very effective at responding to cortisol and down-regulating stress responses via the negative feedback loop described in Figure 12.1.

developed into 'high-stress' adults. Cortisol levels were increased in these mice, especially in response to mild stress, just like the relatively neglected rats.

The researchers working on the mice studied the arginine vasopressin gene. Arginine vasopressin is secreted by the hypo-thalamus, and stimulates secretion from the pituitary. It is shown in Figure 12.1. The stressed-out mice, those that had suffered separation from their mothers in early life, had decreased DNA methylation of the arginine vasopressin gene. This resulted in increased production of arginine vasopressin, which stimulated the stress response[7].

The rat and mouse experimental studies show us two important things. The first is that when early life events lead to adult stress, there is probably more than one gene involved. Both the cortisol receptor gene and the arginine vasopressin gene can contribute to this phenotype in rodents.

Secondly, the studies also show us that a particular class of epigenetic modification is not in itself good or bad. It's where the modification happens that matters. In the rat model, the decreased DNA methylation of the cortisol receptor gene is a 'good' thing. It leads to increased production of this receptor, and a general dampening down of the stress response. In the mouse model, the decreased DNA methylation of the arginine vasopressin gene is a 'bad' thing. It leads to increased expression of this hormone and a stimulation of the stress response.

The decreased DNA methylation of the arginine vasopressin gene in the mouse model occurred through a different route to the one used in the rat hippocampus to activate the cortisol receptor gene.

In the mouse studies, separation from the mother triggered activity of the neurons in the hypothalamus. This set off a signalling cascade that affected the MeCP2 protein. MeCP2 is the protein we met in Chapter 4, which binds to methylated DNA and helps repress gene expression. It's also the gene which is mutated in Rett syndrome, the devastating neurological disorder. Adrian Bird has shown that the MeCP2 protein is incredibly highly expressed in neurons[8].

Normally, MeCP2 protein binds to the methylated DNA at the arginine vasopressin gene. But in the stressed baby mice, the signalling cascade mentioned in the previous paragraph adds a small chemical group called a phosphate to the MeCP2 protein and because of this MeCP2 falls off the arginine vasopressin gene. One of the important roles of MeCP2 is attracting other epigenetic proteins to where it is bound on a gene. These are proteins that all cooperate to add more and more repressive marks to that region of the genome. When the phosphorylated MeCP2 falls off the arginine vasopressin gene, it can no longer recruit these different epigenetic proteins. Because of this, the chromatin loses it repressive marks. Activating modifications get put on instead, such as high levels of histone acetylation. Ultimately, even the DNA methylation is permanently lost.

Amazingly this all happens in the mice in the first ten days after birth. After that, the neurons essentially lose their plasticity. The DNA methylation pattern that's in place at the end of this stage becomes the stable pattern at this location. If the DNA methylation levels are low, this will normally be associated with abnormally high expression of the arginine vasopressin gene. In this way, the early life events trigger epigenetic changes which get effectively 'stuck'. Because of this, the animal continues to be highly stressed, with abnormal hormone production, long after the initial stress has vanished. Indeed, the response continues long after the animal would even normally 'care' about whether or not it has its mother's company. After all, mice are not renowned for hanging about to look after their ageing parents.

In the depths

Researchers are gradually gathering data that suggest some of the changes seen in the rodent models of early stress may be relevant in humans. As mentioned earlier, there are logistical, but more importantly ethical, issues which make it impossible to perform the same kinds of studies in people. Even so, some intriguing correlations are emerging.

The original work in the rat model was carried out by Professor Michael Meaney at McGill University in Montreal. His group subsequently performed some interesting studies on human brain samples from individuals who had, sadly, committed suicide. The group analysed the levels of DNA methylation at the cortisol receptor gene in the hippocampus from these cases. Their data showed that the DNA methylation tended to be higher in the samples from people who had had a history of early child-hood abuse or neglect. By contrast, the DNA methylation levels at this gene were relatively low in the suicide victims who had not had traumatic childhoods[9]. The high DNA methylation levels in the abuse victims would drive down expression of the cortisol receptor gene. This would make the negative feedback loop less efficient and raise the circulating levels of cortisol. This was con-sistent with the findings from the rat work, where the stressed-out animals from the less nurturing mothers had high levels of DNA methylation at the cortisol receptor gene in the hippocampus.

Of course, it isn't just people who have had abusive childhoods who develop mental illnesses. The global figures for depression are startling. The World Health Organisation estimates that over 120 million people worldwide are affected by depression. Depression-related suicides have reached 850,000 per annum and depression is predicted to become the second greatest contributor to the global disease burden by 2020[10].

Effective treatment for depression took a big step forwards in the early 1990s with the licensing by the US Food and Drug Administration of a class of drugs called SSRIs – selective sero-tonin re-uptake inhibitors. Serotonin is a neurotransmitter mol-ecule – it conveys signals between neurons. Serotonin is released in the brain in response to pleasurable stimuli; it's the feel-good molecule that we met in our happy rat babies. The levels of sero-tonin are low in the brains of people suffering from depression. SSRI drugs raise the levels of serotonin in the brain.

It makes sense that drugs that cause an increase in serotonin levels would be useful in treating depression. But there's some-thing odd about their action. The serotonin levels in the brain rise quite quickly when patients are treated with the SSRI drugs. But it usually takes at least four to six weeks before the terrible symp-toms of severe depression begin to lift.

This suggests that there is more to depression than simply a drop in the levels of a single chemical in the brain, which perhaps isn't that surprising. It's very unusual for depression to happen overnight – it's not like coming down with the flu. There's now a reasonable amount of data showing that there are much longer-term changes in the brain as depression develops. These include alterations in the numbers of contacts that neurons make with each other. This in turn is critically dependent on the levels of chemicals called neurotrophic factors[11]. These chemicals support healthy survival and function of brain cells.

Researchers in the depression field have moved away from a simple model based on levels of neurotransmitters and into a more complex network system. This involves sophisticated interactions between neuronal activity and a whole range of other factors. These include stress, production of neurotransmitters, effects on gene expression and longer-term consequences for neurons and how they interact with each other. While this system is in balance, the brain functions healthily. If the system moves out of balance, this complicated network begins to unravel. This moves the brain's biochemistry and function further away from health and closer to dysfunction and disease.

Scientists are beginning to focus their attention in this field on epigenetics, because of its potential to create and sustain long-lasting patterns of gene expression. Rodents are the most common model system for these investigations. Because a mouse or a rat can't tell you how it's feeling, researchers have created certain behavioural tests that are used to model different aspects of human depression.

We all recognise that different people seem to respond to stress in different ways. Some people seem fairly robust. Others can react really badly to the same stressful situation, even developing depression. Mice from different inbred strains are like this as well. Researchers exposed two different strains to mildly stressful stimuli. After the stressful situation, the researchers assessed the behaviour of the mice in some of the tests which mimic certain aspects of human depression. One strain was relatively non-anxious, whereas the other was relatively anxious. These strains were called B6 and BALB, but we'll called them 'chilled' and 'jumpy', respectively, for convenience.

The researchers focused their studies on a region of the brain called the nucleus accumbens. This region plays a role in various emotionally important brain functions. These include aggression, fear, pleasure and reward. The researchers analysed the expression of various neurotrophic factors in the nucleus accumbens. The one that gave the most interesting results was a gene called *Gdnf* (glial cell-derived neurotrophic factor).

Stress caused an increase in expression of the *Gdnf* gene in the chilled mice. In the jumpy strain it caused a decrease in expression of the same gene. Now, different inbred strains of mice can have different DNA codes so the researchers analysed the promoter region, which controls the expression of *Gdnf*. The DNA sequence of the *Gdnf* promoter was identical in the chilled and the jumpy strains. But when the scientists examined the epigenetic modifications in this promoter, they found a difference. The histones of the jumpy mice had fewer acetyl groups than the histones of the chilled mice. As we've seen, low levels of histone acetylation are associated with low levels of gene expression, so this tied up well with the decreased *Gdnf* expression in the jumpy mice.

This led the scientists to wonder what had happened in the neurons of the nucleus accumbens. Why had the levels of histone acetylation dropped at the *Gdnf* gene in the jumpy mice? The scientists examined the levels of the enzymes that add or remove acetyl groups from histones. They found only one difference between the two strains of mice. A specific histone deacetylase (member of the class of proteins which removes acetyl groups) called Hdac2 was much more highly expressed in the neurons of the jumpy mice[12], compared with the chilled out mice.

Other researchers tested mice in a different model of depression, called social defeat. In these experiments, mice are basically humiliated. They're put in an environment where they can't get away from a bigger, scarier mouse, although they are removed before they come to any physical harm. Some mice find this really stressful; others seem to brush it off.

In the experiments adult mice underwent ten days of social defeat. At the end of this they were classified as either susceptible or resistant, depending on how well they bounced back from the experience. Two weeks later the mice were examined. The resistant mice had normal levels of corticotrophin-releasing hormone. This is the chemical released by the hypothalamus. It's the one which ultimately stimulates the production of cortisol, the stress hormone. The susceptible mice had high levels of corticotrophin-releasing hormone and low levels of DNA methylation at the promoter of this gene. This was consistent with the high levels of expression from this gene. They also had low levels of Hdac2, and high levels of histone acetylation, which again fits with over-expression of the corticotrophin-releasing hormone[13].

It might seem odd that in one model system Hdac2 levels went up in the susceptible mice, whereas in another they went down. But it's important with all these epigenetic events to remember that context is everything. There isn't just one way in which Hdac2 levels (or those of any other epigenetic gene, for that matter) are controlled. The control will depend on the region of the brain and the precise signalling pathways that are activated in response to a stimulus.

The drugs do work

There's more evidence supporting a significant role for epigenetics in responses to stress. The naturally jumpy B6 mice were the ones with the increased expression of Hdac2 in the nucleus accumbens, and decreased expression from the *Gdnf* gene. We can treat these mice with SAHA, the histone deacetylase inhibitor. SAHA treatment leads to increased acetylation of the *Gdnf* promoter. This is associated with increased expression of the *Gdnf* gene. The crucial finding is that the treated mice stop being jumpy and become chilled instead[14] – changing the histone acetylation levels of the gene changed the mouse's behaviour. This supports the idea that histone acetylation is really important in modulating the responses of these mice to stress.

One of the tests used to investigate how depressed the mice become in response to stress is called the sucrose-preference test. Normal happy mice love sugared water, but when they are depressed they aren't so interested in it. This decreased response to a pleasant stimulus is called anhedonia. It seems to be one of the best surrogate markers in animals for human depression[15]. Most people who have been severely depressed talk about losing interest in all the things that used to make life joyful before they became ill. When the stressed mice were treated with SSRI anti-depressants, their interest in the sugared water gradually increased. But when they were treated with SAHA, the HDAC inhibitor, they regained their interest in their favourite drink much faster[16].

It's not just in the jumpy or chilled mice that histone deacetylase inhibitors can change animal behaviour. It's also relevant to the baby rats who don't get much maternal licking and grooming. These are the ones that normally grow up to be chronically stressed, with over-activation of the cortisol production pathway. If these 'unloved' animals are treated with TSA, the first histone deacetylase inhibitor to be identified, they grow up much less stressed. They react much more like the animals who received lots of maternal care. The levels of DNA methylation at the cortisol receptor gene in the hippocampus go down, increasing expression of the receptor and improving the sensitivity of the all-important negative feedback loop. This is presumed to be because of cross-talk between the histone acetylation and DNA methylation pathways[17].

In the social defeat model in mice, the susceptible animals were treated with an SSRI anti-depressant drug. After three weeks of treatment, their behaviour was much more like that of the resilient mice. But treatment with this anti-depressant drug didn't just result in increased levels of serotonin in the brain. The anti-depressant treatment also led to increased DNA methylation at the promoter of the corticotrophin-releasing hormone.

These studies are all very consistent with a model where there is cross-talk between the immediate signals from the neurotransmitters, and the longer-term effects on cell function mediated by epigenetic enzymes. When depressed patients are treated with SSRI drugs, the serotonin levels in the brain begin to rise, and signal more strongly to the neurons. The animal work described in the last paragraph suggests that it takes a few weeks for these signals to trigger all the pathways that ultimately result in the altered pattern of epigenetic modifications in the cells. This stage is essential for restoring normal brain function.

Epigenetics is also a reasonable hypothesis to explain another interesting but distressing feature of severe depression. If you have suffered from depression once, you are at a significantly higher risk than the general population of suffering from it again at some time in the future. It's likely that some epigenetic modifications are exceptionally difficult to reverse, and leave the neurons primed to be more vulnerable to another bout.

The jury's out

So far, so good. Everything looks very consistent with our theory about life experiences having sustained and long-lasting effects on behaviour, through epigenetics. And yet, here's the thing: this whole area, sometimes called neuro-epigenetics, is probably the most scientifically contentious field in the whole of epigenetic research.

To get a sense of just how controversial, consider this. We've met Professor Adrian Bird in this book before. He is acknowledged as the father of the DNA methylation field. Another scientist with a very strong reputation in the science behind DNA methylation is Professor Tim Bestor from Columbia University Medical Center in New York. Adrian and Tim are about the same age, of similar physical type, and both are thoughtful and low key in conversation. And they seem to disagree on almost every issue in DNA methylation. Go to any conference where they are both scheduled in the same session and you are guaranteed to witness inspiring and impassioned debate between the two men. Yet the one thing they both seem to agree on publicly is their scepticism about some of the reports in the neuro-epigenetics field[18].

There are three reasons why they, and many of their colleagues, are so sceptical. The first is that many of the epigenetic changes that have been observed are relatively small. The sceptics are unconvinced that such small molecular changes could lead to such pronounced phenotypes. They argue that just because the changes are present, it doesn't mean they're necessarily having a functional effect. They worry that the alterations in epigenetic modifications are simply correlative, not causative.

The scientists who have been investigating the behavioural responses in the different rodent systems counter this by arguing that molecular biologists are too used to quite artificial experimental models, where they can study extensive molecular changes with very on-or-off read-outs. The behaviourists suspect that this has left molecular biologists relatively inexperienced at interpreting real-world experiments, where the read-outs tend to be more 'fuzzy' and prone to greater experimental variation.

The second reason for scepticism lies in the very localised nature of the epigenetic changes. Infant stress affects specific regions of the brain, such as the nucleus accumbens, and not other areas. Epigenetic marks are only altered at some genes and not others. This seems less of a reason for scepticism. Although we refer to 'the brain', there are lots of highly specialised centres and regions within this organ, the product of hundreds of millions of years of evolution. Somehow, all these separate regions are generated and maintained during development and beyond, and thus are clearly able to respond differently to stimuli. This is also the case for all our genes, in all our tissues. It's true that we don't really know how epigenetic modifications can be targeted so precisely, or how the signalling from chemicals like neurotransmitters leads to this targeting. But we know that similarly specific events occur during normal development – so why not during abnormal periods of stress or other environmental disturbances? Just because we don't know the mechanism for something, it doesn't mean it doesn't happen. After all, John Gurdon didn't know how adult nuclei were reprogrammed by the cytoplasm of eggs, but that didn't mean his experimental findings were invalid.

The third reason for scepticism is possibly the most important and it relates to DNA methylation itself. DNA methylation at the target genes in the brain is established very early, possibly pre-natally but certainly within one day of birth, in rodents. What this means is that the baby mice or baby rats in the experiments all started life with a certain baseline pattern of DNA methylation at their cortisol receptor gene in the hippocampus. The DNA methylation levels at this promoter alter in the first week of life, depending on the amount of licking and grooming the rats receive. As we saw, the DNA methylation levels are higher in the neglected mice than in the loved ones. But that's not because the DNA methylation has gone up in the neglected mice. It's because DNA methylation has gone *down* in the ones that were licked and groomed the most. The same is also true at the arginine vasopressin gene in the baby mice removed from their mothers. It's also true for the corticotrophin-releasing hormone gene in the adult mice that were susceptible to social defeat.

So, in every case, what the scientists observed was decreased DNA methylation in response to a stimulus. And that's where, molecularly, the problem lies, because no-one knows how this happens. In Chapter 4 we saw how copying of methylated DNA results in one strand that contains methyl groups and one that doesn't. The DNMT1 enzyme moves along the newly synthesised strand and adds methyl groups to restore the methylation pattern, using the original strand as a template. We could speculate that in our experimental animals, there was less DNMT1 enzyme present and so the methylation levels at the gene dropped. This is referred to as passive DNA demethylation.

The problem is that this can't work in neurons. Neurons are terminally differentiated – they are right at the bottom of Waddington's landscape, and cannot divide. Because they don't divide, neurons don't copy their DNA. There's no reason for them to do so. As a result, they can't lose their DNA methylation by the method described in Chapter 4.

One possibility is that maybe neurons simply remove the methyl group from DNA. After all, histone deacetylases remove acetyl groups from histones. But the methyl group on DNA is different. In chemical terms, histone acetylation is a bit like adding a small Lego brick onto a larger Lego brick. It's pretty easy to take the two bricks apart again. DNA methylation isn't like that. It's more like having two Lego bricks and using superglue to stick them together.

The chemical bond between a methyl group and the cytosine in DNA is so strong that for many years it was considered completely irreversible. In 2000, a group from the Max Planck Institute in Berlin demonstrated that this couldn't be the case. They showed that in mammals the paternal genome undergoes extensive DNA demethylation, during very early development. We came across this in Chapters 7 and 8. What we glossed over at the time was that this demethylation happens before the zygote starts to divide. In other words, the DNA methylation was removed without any DNA replication[19]. This is referred to as active DNA demethylation.

This means there is a precedent for removing DNA methylation in non-dividing cells. Perhaps there's a similar mechanism in neurons. There's still a lot of debate about how DNA methylation is actively removed, even in the well-established events in early development. There's even less consensus about how it takes place in neurons. One of the reasons this has been so hard to investigate is that active DNA demethylation may involve a lot of different proteins, carrying out a number of steps one after another. This makes it very difficult to recreate the process in a lab, which is the gold standard for these kinds of investigations.

Silencing the silencer

As we've seen repeatedly, scientific research often throws up some very unexpected findings and so it happened here. While many people in epigenetics were looking for an enzyme that removed DNA methylation, one group discovered enzymes that added something extra to methylated DNA. This is shown in Figure 12.3. Very surprisingly, this has turned out to have many of the same consequences as demethylating the nucleic acid.

A small molecule called hydroxyl, consisting of one oxygen atom and one hydrogen atom, is added to the methyl group, to

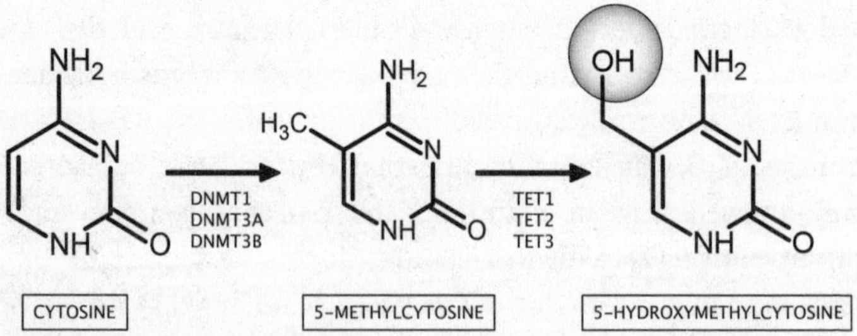

Figure 12.3   Conversion of 5-methylcytosine to 5-hydroxymethylcytosine. C: carbon; H: hydrogen; N: nitrogen; O: oxygen. For simplicity, some carbon atoms have not been explicitly shown, but are present where there is a junction of two lines.

create 5-hydroxymethylcytosine. This reaction is carried out by enzymes called TET1, TET2 or TET3[20].

This is highly relevant to the question of DNA demethylation, because it's the effects of DNA methylation that make this change important. Methylation of cytosine affects gene expression because methylated cytosine binds certain proteins, such as MeCP2. MeCP2 acts with other proteins to repress gene expression and to recruit other repressive modifications like histone deacetylation. When an enzyme such as TET1 adds the hydroxyl group to the methylcytosine to form the 5-hydroxymethylcytosine molecule, it changes the shape of the epigenetic modification. If a methylated cytosine is like a grape on a tennis ball, the 5-hydroxymethylcytosine is like a bean stuck to a grape stuck to a tennis ball. Because of this change in shape, the MeCP2 protein can't bind to the modified DNA any more. The cell therefore 'reads' 5-hydroxymethylcytosine in the same way as it reads unmethylated DNA.

Many of the techniques used until very recently looked for the presence of DNA methylation. They often couldn't distinguish between unmethylated DNA and 5-hydroxymethylated DNA. This means that many of the papers which refer to decreased DNA methylation may actually have been detecting increased 5-hydroxymethylation without knowing it. It's currently unproven, but it may be that instead of actually demethylating DNA, as reported in some of the behavioural studies, neurons really convert 5-methylcytosine to 5-hydroxymethylcytosine. The techniques for studying 5-hydroxymethylcytosine are still under development but we do know that neurons contain higher levels of this chemical than any other cell type[21].

Remember, remember

Despite these controversies, research is continuing into the importance of epigenetic modifications in brain function. One area that is attracting a lot of attention is the field of memory. Memory is an incredibly complex phenomenon. Both the hippocampus and a region of the brain called the cortex are involved in memory, but in different ways. The hippocampus is mainly involved in consolidating memories, as our brains decide what we are going to remember. The hippocampus is fairly plastic in the way that it operates, and this seems to be associated with transient changes in DNA methylation, again through fairly uncharacterised mechanisms. The cortex is used for longer-term storage of memories. When memories are stored in the cortex, there are prolonged changes in DNA methylation.

The cortex is like a hard drive on a computer with gigabytes of storage. The hippocampus is more like the RAM (random access memory) chip, where data are temporarily processed before being deleted, or transferred to the hard drive for permanent storage. Our brain separates out different functions to selected cell populations in different anatomical regions. This is why memory loss is rarely all-encompassing. Depending on the clinical condition, for example, either one of short-term or long-term memory may be relatively lost or remain relatively intact. It makes a lot of sense for these different functions to be separated in our brains. Just try to imagine life if we remembered everything that ever happened – the phone number that we dialled only once, every word a dull stranger said to us on a train, or the canteen menu from a wet Wednesday three years ago.

The complexity of our memory systems is one of the reasons why it is quite a difficult area to study, because it can be difficult to set up experiments where we are absolutely sure which aspects of memory our experimental techniques are actually addressing. But one thing we know for sure is that memory involves long-term changes in gene expression, and in the way neurons make connections with one another. And that again leads to the hypothesis that epigenetic mechanisms may play a role.

In mammals, both DNA methylation and histone modifications play a role in memory and learning. Rodent studies have shown that these changes may be targeted to very specific genes in discrete regions of the brain, as we have come to expect. For example, the DNA methyltransferase proteins DNMT3A and DNMT3B increase in expression in the adult rat hippocampus in a particular learning and memory model. Conversely, treating these rats with a DNA methyltransferase inhibitor such as 5-azacytidine blocks memory formation and affects both the hippocampus and the cortex[22].

A particular histone acetyltransferase (protein which adds acetyl groups to histones) gene is mutated in a human disorder called Rubinstein-Taybi syndrome. Mental retardation is a frequent symptom in this disease. Mice with a mutant version of this gene also have low levels of histone acetylation in the hippocampus, as we would predict. They also have major problems in long-term memory processing in the hippocampus[23]. When these mice were treated with SAHA, the histone deacetylase inhibitor, acetylation levels in the hippocampus went up, and the memory problems improved[24].

SAHA can inhibit many different histone deacetylases, but in the brain some of its targets seem to be more important than others. The two most highly expressed enzymes of this class are HDAC1 and HDAC2. These differ in the ways they are expressed in the brain. HDAC1 is predominantly expressed in neural stem cells, and in a supportive, protective population of non-neurons called glial cells. HDAC2 is predominantly expressed in neuronal cells[25], so it's unsurprising that this is the histone deacetylase that is most important in learning and memory.

Mice whose neurons over-express Hdac2 have poor long-term memory, even though their short-term memory is fine. Mice whose neurons don't express any Hdac2 have excellent memories. These data show us that Hdac2 has a negative effect on memory storage. The neurons which over-expressed Hdac2 formed far fewer connections than normal, whereas the opposite was true for the neurons lacking Hdac2. This supports our model of epigenetically-driven changes in gene expression ultimately altering complex networks in the brain. SAHA improves memory in the mice that over-express Hdac2, presumably by dampening down its effects on histone acetylation and gene expression. SAHA also improves memory in normal mice[26].

In fact, increased acetylation levels in the brain seem to be consistently associated with improved memory. Learning and memory both improved in mice kept in conditions known as environmentally enriched. This is a fancy way of saying they had access to two running wheels and the inside of a toilet roll. The histone acetylation levels in the hippocampus and cortex were increased in the mice in the more entertaining surroundings. Even in these mice, the histone acetylation levels and memory skills improved yet further if they were treated with SAHA[27].

We can see a consistent trend emerging. In various different model systems, learning and memory improve when animals are treated with DNA methyltransferase inhibitors, and especially with histone deacetylase inhibitors. As we saw in the last chapter, there are drugs licensed in both these classes, such as 5-aza-cytidine and SAHA, respectively. It's very tempting to speculate about taking these anti-cancer drugs and using them in conditions where memory loss is a major clinical problem, such as Alzheimer's disease. Perhaps we might even use them as general memory enhancers in the wider population.

Unfortunately, there are substantial difficulties in doing this. These drugs have side-effects which can include severe fatigue, nausea and a higher risk of infections. These side-effects are considered acceptable if the alternative is an inevitable and fairly near-term death from cancer. But they might be considered less acceptable for treating the early stages of dementia, when the patient still has a relatively reasonable quality of life. And they would certainly be unacceptable for the general population.

There is an additional problem. Most of these drugs are really bad at getting into the brain. In many of the rodent experiments, the drugs were administered directly into the brain, and often into very defined regions such as the hippocampus. This isn't a realistic treatment method for humans.

There are a few histone deacetylase inhibitors that do get into the brain. A drug called sodium valproate has been used for decades to treat epilepsy, and clearly must be getting into the brain in order to do this. In recent years, we have realised that this compound is also a histone deacetylase inhibitor. This would be extremely encouraging for trying to use epigenetic drugs in Alzheimer's disease but unfortunately, sodium valproate only inhibits histone deacetylases very weakly. All the animal data on learning and memory have shown that stronger inhibitors work much better than weak ones at reversing these deficits.

It's not just in disorders like Alzheimer's disease that epigenetic therapies could be useful if we manage to develop suitable drugs. Between 5 and 10 per cent of regular users of cocaine become addicted to the drug, suffering from uncontrollable cravings for this stimulant. A similar phenomenon occurs in rodents, if animals are allowed unlimited access to the drug. Addiction to stimulants such as cocaine is a classic example of inappropriate adaptations by memory and reward circuits in the brain. These maladaptations are regulated by long-lasting changes in gene expression. Changes in DNA methylation, and in how methylation is read by MeCP2, underpin this addiction. This happens via a set of poorly understood interactions which include signalling factors, DNA and histone modifying enzymes and readers, and miRNAs. Related pathways also underpin addiction to amphetamines[28,29].

If we return to the starting point of this chapter, it's clear that there's a major need to stop children who have suffered early trauma from developing into adults with a substantially higher than normal risk of mental illness. It's very appealing to think we might be able to use epigenetic drug therapies to improve their life chances. Unfortunately, one of the problems in designing therapies for children who have been abused or neglected is that it's actually pretty difficult to identify those who will be permanently damaged as adults, and those who will have healthy, happy and fulfilled lives. There are enormous ethical dilemmas around giving drugs to children, when we can't be sure if an individual child actually needs the treatment. In addition, clinical trials to determine if the drugs actually do any good would need to last for decades, which makes them economically almost a non-starter for any pharmaceutical company.

But we mustn't end on too negative a note. Here's a great story about an epigenetic event and behaviour. There is a gene called *Grb10* that is involved in various signalling pathways. It's an imprinted gene, and the brain only expresses the paternally inherited copy. If we switch off this paternal copy, the mouse can't produce any Grb10 protein, and the animals develop a very odd phenotype. They nibble off the face fur and whiskers of other mice in the same cage. This is a sort of aggressive grooming, a bit like a pecking order in chickens. In addition, if faced with a big mouse that they don't know, the *Grb10* mutant mice don't back away – they stand their ground[30].

Switching off *Grb10* in the brain results in what might sound like a rather impressive, kick-ass kind of a mouse. It maybe even seems odd that this gene is normally switched on in the brain. Wouldn't mice that switched off *Grb10* be the butchest, most successful mice? Actually, it's more likely that they'd be the mice most likely to get themselves beaten up. There are a lot of mice in the world, and they encounter each other pretty frequently. It pays to recognise when you are out-gunned.

When the *Grb10* gene is switched off in the brain, it's like a bad Friday night for the mouse. Let's put this in human terms so we can see why. You're down the pub when a person twice your size and all muscle knocks against you and you spill your pint.

When this gene is switched off, it's as if you have a friend next to you who says, 'Go on, you can take him/her, don't wimp out.' We all know how badly those scenarios tend to play out. So let's end this chapter by raising a cheer for imprinted *Grb10*, the gene that likes to say, 'Leave it mate, it's not worth it.'

Chapter 13

The Downhill Slope

I guess I don't so much mind being old, as I mind being fat and old.
Benjamin Franklin

Time moves forward, we age. It's inevitable. And as we get older, our bodies change. Once we're past our mid-thirties most of us would agree that it gets harder and harder to sustain the same level of physical performance. It doesn't matter if it's how fast we can run, how far we can cycle before needing to stop for a break, or how quickly we recover from a big night out. The older we get, the harder everything seems to become. We develop new aches and pains, and succumb more easily to annoying little infections.

Ageing is something we are good at recognising in the people around us. Even quite small children can tell the difference between the young and the very old, even if they are a bit hazy on everyone in the middle. Adults can easily tell the difference between a 20-year-old and a 40-something individual, or between two people who are 40 and 65.

We can categorise individuals instinctively into approximate age groups not because they give off an intrinsic radio signal about the number of years they have been on earth, but because of the physical signs of ageing. These include the loss of fat beneath the skin, making our features more drawn and less 'fresh-faced'. There are the wrinkles, the fall in muscle tone, that slight curvature to the spine.

The growth of the cosmetic surgery industry appears to be relentless and shows how desperate we can be to fight the symptoms of ageing. Figures released in 2010 showed that in the top

25 countries covered in a survey by the International Society of Aesthetic Plastic Surgery, there were over eight and a half million surgical procedures carried out in 2009, and about the same number of non-surgical procedures, such as Botox and dermo-abrasion. The United States topped the list, with Brazil and China fighting for second place[1].

As a society, we don't seem to mind really about the number of years we've been alive, but we dislike intensely the physical decline that accompanies them. It's not just the trivial stuff either. One of the greatest risk factors for developing cancer is simply being old. The same is true for conditions such as Alzheimer's disease and stroke.

Most breakthroughs in human healthcare up until now have improved both longevity and quality of life. That's partly because many major advances targeted early childhood deaths. Vaccination against serious diseases such as polio, for example, has hugely improved both childhood mortality figures (fewer children dying) and morbidity in terms of quality of life for survivors (fewer children permanently disabled as a result of polio).

There is a growing debate around the issue sometimes known as human life extension, which deals with extending the far end of life, old age. Human life extension refers to the concept that we can use interventions so that individuals will live to a greater age. But this takes us into difficult territory, both socially and scientifically. To understand why, it's important to establish what ageing really is, and why it is so much more than just being alive for a long time.

One useful definition of ageing is 'the progressive functional decline of tissue function that eventually results in mortality'[2]. It's this functional decline that is the most depressing aspect of ageing for most people, rather than the final destination.

Generally speaking, most of us recognise the importance of this quality of life issue. For example, in a survey of 605 Australian adults in 2010, about half said they would not take an anti-ageing pill if one were developed. The rationale behind their choice was based around quality of life. These respondents didn't believe such a pill would prolong healthy life. Simply living for longer wasn't attractive, if this was associated with increasing ill-health and disability. These respondents did not wish to prolong their own lives, unless this was associated with improved health in later years[3].

There are thus two separate aspects to any scientific discussion of ageing. These are lifespan itself, and the control of late-onset disorders associated with ageing. What isn't clear is the degree to which it is possible or reasonable to separate the two, at least in humans.

Epigenetics definitely has a role to play in ageing. It's not the only factor that's important, but it is significant. This field of epigenetics and ageing has also led to one of the most acrimonious disputes in the pharmaceutical sector in recent years, as we'll see towards the end of this chapter.

We have to ask why our cells malfunction as we get older, leaving us more at risk of illnesses that include cancer, type 2 diabetes, cardiovascular disease and dementia, amongst a host of other conditions. One reason is because the DNA script in the cells of our body begins to change for the worse. It accumulates random alterations in sequence. These are somatic mutations, which affect the tissue cells of the body, but not the germline. Many cancers have changes in the DNA sequence, often caused by quite large rearrangements between chromosomes, where genetic material is swapped from one chromosome to another.

Guilt by association

But as we've seen, our cells contain multiple mechanisms for keeping the DNA blueprint as intact as possible. Wherever possible, a cell's default setting is to maintain the genome in its original state, as much as it can. But the epigenome is different. By its very nature, this is more flexible and plastic than the genome. Because of this, it is probably not surprising that epigenetic modifications change as animals age. The epigenome may eventually turn out to be far more prone to changes with age than the genome, because the epigenome is more naturally variable than the genome anyway.

We met some examples of this in Chapter 5, where we discussed how genetically identical twins become less identical epigenetically as they age. The issue of how the epigenome changes as we age has been examined even more directly. Researchers studied two large groups of people from Iceland and from Utah, who have been part of on-going long-term population studies. DNA was prepared from blood samples that had been taken from these people between eleven and sixteen years apart. Blood contains red and white blood cells. The red blood cells carry oxygen around the body, and are essentially just little bags of haemoglobin. The white blood cells are the cells that generate immune responses to infections. These cells retain their nuclei and contain DNA.

The researchers found that the overall DNA methylation levels in the white blood cells of some of these individuals changed over time. The change wasn't always the same. In some individuals, the DNA methylation levels went up with age, in others they dropped. The direction of change seemed to run in families. This may mean that the age-related change in DNA methylation was genetically influenced, or affected by shared environmental factors in a family. The scientists also looked in detail at methylation at over 1,500 specific CpG sites in the genome. These sites were mainly associated with protein-coding genes. They found the same trends at these specific sites as they had seen when looking at overall DNA methylation levels. In some individuals, site-specific DNA methylation was increased whereas in others it fell. DNA methylation levels were increased or decreased by at least 20 per cent in around one tenth of the people in the study.

The authors stated in their conclusion that 'these data support the idea of age-related loss of normal epigenetic patterns as a mechanism for late onset of common human diseases'[4]. It's true that the data are consistent with this model of epigenetic mechanisms leading to late onset disease, but there are limitations, which we should bear in mind.

In particular, these types of studies highlight important correlations between epigenetic change and diseases of old age, but they don't prove that one event causes the other. Deaths through drowning are most common when sales of suntan lotion are highest. From this one could infer that sun tan lotions have some effect on people that makes them more likely to drown. The reality of course is that sales of suntan lotion rise during hot weather, which is also when people are most likely to go swimming. The more people who swim, the greater the number who will drown, on average. There is a correlation between the two factors we have monitored (sales of sun block and deaths by drowning) but this isn't because one factor causes the other.

So, although we know that epigenetic modifications change over time, this doesn't prove that these alterations cause the illnesses and degeneration associated with old age. In theory, the changes could just be random variations with no functional consequences. They could just be changes in the epigenetic background noise in a cell. In many cases, we don't even yet know whether the altered patterns of epigenetic modifications lead to changes in gene expression. Addressing this question is hugely challenging, and particularly difficult to assess in human populations.

Guilt by more than association

Having said that, there are some epigenetic modifications that are definitely involved in disease initiation or progression. The case for these is strongest in cancer, as we saw in Chapter 11. The evidence includes the epigenetic drugs which can treat certain specific types of cancer. It also includes the substantial amounts of data from experimental systems. These show that altering epigenetic regulation in a cell increases the likelihood of a cell becoming cancerous, or can make an already cancerous cell more aggressive.

One of the areas that we dealt with in Chapter 11 was the increase in DNA methylation that frequently occurs at the promoters of tumour suppressor genes. This increased DNA methylation switches off the expression of the tumour suppressor genes. Oddly enough, this increase in DNA methylation at specific sites is often found against an overall background of decreased DNA methylation in many other areas of the genome in the same cancer cell. This decrease in methylation may be caused by a fall in expression or activity of the maintenance DNA methyltransferase, DNMT1. This decrease in global DNA methylation may also contribute to the development of cancer.

To investigate this, Rudi Jaenisch generated mice which only expressed Dnmt1 protein at about 10 per cent of normal levels in their cells. The levels of DNA methylation in their cells were very low compared with normal mice. In addition to being quite stunted at birth, these *Dnmt1* mutant mice developed aggressive tumours of the immune system (T cell lymphomas) when they were between four and eight months of age. This was associated with rearrangements of certain chromosomes, and especially with an extra copy of chromosome 15 in the cancer cells.

Professor Jaenisch speculated that the low levels of DNA methylation made the chromosomes very unstable and prone to breakages. This put the chromosomes at high risk of joining up in inappropriate ways. It's like snapping a pink stick of rock and a green stick of rock to create four pieces in total. You can join them back together again using melted sugar, to create two full-length items of tooth-rotting confectionery. But if you do this in the dark, you may find that sometimes you have created 'hybrid' rock sticks, where one part is pink and the other is green.

The end result of increased chromosome instability in Rudi Jaensich's mice was abnormal gene expression. This in turn led to too much proliferation of highly invasive and aggressive cells, resulting in cancer[5,6]. These data are one of the reasons why DNMT inhibitors are unlikely to be used as drugs in anything other than cancer. The fear is that the drugs would cause decreased DNA methylation in normal cells, which might pre-dispose some cell types towards cancer.

These data suggest that the DNA methylation level per se is not the critical issue. What matters is *where* the changes in DNA methylation take place in the genome.

The generalised decrease in DNA methylation levels that comes with age has also been reported in other species than humans and mice, ranging from rats to humpback salmon[7]. It's not entirely clear why low levels of DNA methylation are associated with instability of the genome. It may be because high levels of DNA methylation can lead to a very compacted DNA structure, which may be more structurally stable. After all, it's easy to snip through a single extended wire with a pair of cutters, but much harder if that wire has been squashed down into a dense knot of metal.

It's important to appreciate just how much effort cells put into looking after their chromosomes. If a chromosome breaks, the cell will repair the break if it can. If it can't, the cell may trigger an auto-destruct mechanism, essentially committing cellular suicide. That's because damaged chromosomes can be dangerous. It's better to kill one cell, than for it to survive with damaged genetic material. For instance, imagine one copy of chromosome 9 and one copy of chromosome 22 break in the same cell. They could get repaired properly, but sometimes the repair goes wrong and part of chromosome 9 joins up with part of chromosome 22.

This rearrangement of chromosomes 9 and 22 actually happens relatively frequently in cells of the immune system. In fact it happens so often that this 9:22 hybrid has a specific name. It's called the Philadelphia chromosome, after the city where it was first described. Ninety-five per cent of people who have a form of cancer called chronic myeloid leukaemia have the Philadelphia chromosome in their cancer cells. This abnormal chromosome causes this cancer in the cells of the immune system because of where the breaking and rejoining happen in the genome. The fusion of the two chromosome regions results in the creation of a hybrid gene called *Bcr-Abl*. The protein encoded by this hybrid gene drives cell proliferation forwards very aggressively.

Our cells have therefore developed very sophisticated and fast-acting pathways to repair chromosome breaks as rapidly as possible, in order to prevent these sorts of fusions. To do this, our cells must be able to recognise loose ends of DNA. These are created when a chromosome breaks in two.

But there's a problem. Every chromosome in our cell quite naturally has two loose ends of DNA, one at each end. Something must stop the DNA repair machinery from thinking these ends need to be repaired. That something is a specialised structure called the telomere. There is a telomere at each end of every chromosome, making a total of 92 telomeres per cell in humans. They stop the DNA repair machinery from targeting the ends of chromosomes.

The tail ends

Telomeres play a critical role in control of ageing. The more a cell divides, the smaller its telomeres become. Essentially, as we age, the telomeres get shorter. Eventually, they get so small that they don't function properly anymore. The cells stop dividing and may even activate their self-destruct mechanisms. The only cells where this is different are the germ cells that ultimately give rise to eggs or sperm. In these cells the telomeres always stay long, so the next generation isn't short-changed when it comes to longevity. In 2009, the Nobel Prize in Physiology or Medicine was awarded to Elizabeth Blackburn, Carol Greider and Jack Szostak for their work showing how telomeres function.

Since telomeres are so important in ageing, it makes sense to consider how they interact with the epigenetic system. The DNA of vertebrate telomeres consists of hundreds of repeats of the sequence TTAGGG. There are no genes at the telomere. We can also see from the sequence that there are no CpG motifs at the telomeres, so there can't be any DNA methylation. If there are any epigenetic effects that make a difference at the telomeres they will therefore have to be based on histone modifications.

In between the telomeres and the main parts of the chromosome are stretches of DNA referred to as sub-telomeric regions. These contain lots of runs of repetitive DNA. These repeats are less restricted in sequence than the telomeres. The sub-telomeric regions contain a low frequency of genes. They contain some CpG motifs so these regions can be modified by DNA methylation, in addition to histone modifications.

The types of epigenetic modifications normally found at telomeres and the sub-telomeric regions are the ones that are highly repressive. Because there are so few genes in these regions anyway, these modifications probably aren't used to switch off individual genes. Instead, these repressive epigenetic modifications are probably involved in 'squashing down' the ends of the chromosomes. The epigenetic modifications attract proteins that coat the ends of the chromosomes, and help them to stay as tightly coiled up, and as dense and inaccessible as possible. It's a little like covering the ends of a pipe in insulation.

It's potentially a problem for a cell that all its telomeres have the same DNA sequence, because identical sequences in a nucleus tend to find and bind to one another. Such close proximity creates a big risk that the ends of different chromosomes will link up, especially if they get damaged and opened up. This can lead to all sorts of errors as the cell struggles to sort out chains of chromosomes, and may result in 'mixed-up' chromosomes similar to the one that causes chronic myeloid leukaemia. By coating the telomeres with repressive modifications that make the ends of the chromosomes really densely packed, there's less chance that different chromosomes will join up inappropriately.

The cell is, however, stuck with a dilemma, as shown in Figure 13.1.

If the telomeres get too short, the cell tends to shut down. But if the telomeres get too long, there's an increased risk of different chromosomes linking up, and creating new cancer-promoting genes. Cell shut-down is probably a defence mechanism that has evolved to minimise the risk of creating new cancer-inducing genes. This is one of the reasons why it's likely to be very difficult to create drugs that increase longevity without increasing the risk of cancer as well.

What happens when we create new pluripotent cells? This could be through somatic cell nuclear transfer, as we saw in Chapter 1, or through creation of iPS cells, as we saw in Chapter 2. We may use these techniques to create cloned non-human animals, or human stem cells to treat degenerative diseases. In both cases, we want to create cells with normal longevity. After all, there is little point creating a new prize stallion, or cells to implant into the pancreas of a teenager with diabetes, if the horse or the cells die of telomere 'old age' within a short time.

That means we need to create cells with telomeres that are about the same length as the ones in normal embryos. In nature, this occurs because the chromosomes in the germline are protected from telomere shortening. But if we are generating pluripotent cells from relatively adult cells, we are dealing with nuclei where

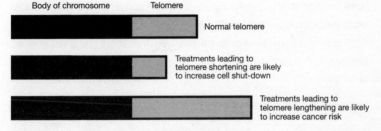

Figure 13.1   Both abnormal shortening and lengthening of telomeres have potentially deleterious consequences for cells.

the telomeres are already likely to be relatively short, because the 'starter' cells were taken from adults, whose chromosomes are getting shorter with age.

Luckily, something unusual happens when we create pluripotent cells experimentally. When iPS cells are created, they switch on expression of a gene called telomerase. Telomerase normally keeps telomeres at a healthy long length. However, as we get older, the telomerase activity in our cells starts to drop. It's important to switch on telomerase in iPS cells, or the cells would have very short telomeres and wouldn't create very many generations of daughter cells. The Yamanaka factors induce the expression of high levels of telomerase in iPS cells.

But we can't use telomerase to try to reverse or slow human ageing. Even if we could introduce this enzyme into cells, perhaps by using gene therapy, the chances of inducing cancers would be too great. The telomere system is finely balanced, and so is the trade-off between ageing and cancer.

Both histone deacetylase inhibitors and DNA methyltransferase inhibitors improve the efficiency of the Yamanaka factors. This might be partly because these compounds help to remove some of the repressive modifications at the telomeres and sub-telomeric regions. This may make it easier for telomerase to build up the telomeres as the cells are reprogrammed.

The interaction of epigenetic modifications with the telomere system takes us a little further away from a simple correlation between epigenetics and ageing. It moves us closer to a model where we can start to develop confidence that epigenetic mechanisms may actually play a causative role in at least some aspects of ageing.

Is your beer getting old?

To investigate this more fully, scientists have made extensive use of an organism we all encounter every day of our lives, whenever we eat a slice of bread or drink a bottle of beer. The scientific term for this model organism is *Saccharomyces cerevisiae*, but we generally know it by its more common name of brewer's yeast. We'll stick with yeast, for short.

Although yeast is a simple one-celled organism, it is actually very like us in some really fundamental ways. It has a nucleus in its cells (bacteria don't have this) and contains many of the same proteins and biochemical pathways as higher organisms such as mammals.

Because yeast are such simple organisms, they're very easy to work with experimentally. A yeast cell (mother) can generate new cells (daughters) in a relatively straightforward way. The mother cell copies its DNA. A new cell buds off from the side of the mother cell. This daughter cell contains the correct amount of DNA, and drifts off to act as a completely independent new single-celled organism. Yeast divide to form new cells really quickly, meaning experiments can be run in a few weeks rather than taking the months or years that are required for some higher organisms, and especially mammals. Yeast can be grown either in a liquid soup, or plated out onto a Petri dish, making them very easy to handle. It's also fairly straightforward to create mutations in interesting genes.

Yeast have a specific feature that has made them one of the favourite model systems of epigeneticists. Yeast never methylate their DNA, so all epigenetic effects *must* be caused by histone modifications. There's also another helpful feature of yeast. Each time a yeast mother cell gives rise to a daughter cell, the bud leaves a scar on the mother. This makes it really easy to work out how many times a cell has divided. There are two types of ageing in yeast and these each have parallels to human ageing, as shown in Figure 13.2.

Most of the emphasis in ageing research has been on replicative ageing, and trying to understand why cells lose their ability to divide. Replicative ageing in mammals is clearly related to some

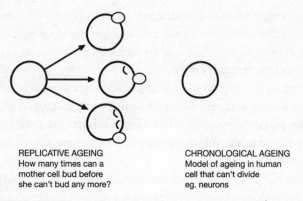

REPLICATIVE AGEING
How many times can a
mother cell bud before
she can't bud any more?

CHRONOLOGICAL AGEING
Model of ageing in human
cell that can't divide
eg. neurons

Figure 13.2 The two models of ageing in yeast, relevant for dividing and non-dividing cells.

obvious symptoms of getting older. For example, skeletal muscle contains specialised stem cells called satellite cells. These can only divide a certain number of times. Once they are exhausted, you can't create new muscle fibres.

Substantial progress has been made in understanding replicative ageing in yeast. One of the key enzymes in controlling this process is called Sir2 and it's an epigenetic protein. It affects replicative ageing in yeast through two pathways. One seems to be specific to yeast, but the other is found in numerous species right through the evolutionary tree, all the way up to humans.

Sir2 is a histone deacetylase. Mutant yeast that over-express Sir2 have a replicative lifespan that is at least 30 per cent longer than normal[8]. Conversely, yeast that don't express Sir2 have a reduced lifespan[9], about 50 per cent shorter than usual. In 2009, Professor Shelley Berger, an incredibly dynamic scientist at the University of Pennsylvania whose group has been very influential in molecular epigenetics, published the results of a really elegant set of genetic and molecular experiments in yeast.

Her research showed that the Sir2 protein influences ageing by taking acetyl groups off histone proteins, and not through any other roles this enzyme might carry out[10]. This was a key experiment, because Sir2, like many histone deacetylases, has rather loose molecular morals. It doesn't just remove acetyl groups from histone proteins. Sir2 will take acetyl groups away from at least 60 other proteins in the cell. Many of these proteins have nothing to do with chromatin or with gene expression. Shelley Berger's work was crucial for demonstrating that Sir2 influences ageing precisely because of its effects on histone proteins. The altered epigenetic pattern on the histones in turn influenced gene expression.

These data, showing that epigenetic modifications of histones really do have a major influence on ageing, gave scientists in this field a big confidence boost that they were on the right track. The importance of Sir2 doesn't seem to be restricted to yeast. If we over-express Sir2 in our favourite worm, *C. elegans*[11], the worm lives longer. Fruit flies that over-expressed Sir2 had up to a 57 per cent increase in lifespan[12]. So, could this gene also be important in human ageing?

There are seven versions of the *Sir2* gene in mammals, called *SIRT1* through to *SIRT7*. Much of the attention in the human field has focused on *SIRT6*, an unusual histone deacetylase. The breakthroughs in this field have come from the laboratory of Katrin Chua, a young Assistant Professor at the Stanford Center on Longevity (and also the sister of Amy Chua who wrote the highly controversial mothering memoir *Battle Hymn of the Tiger Mother*).

Katrin Chua created mice which never expressed any Sirt6 protein, even during their development (they are known as *Sirt6* knockout mice). These animals seemed normal at birth, although they were rather small. But from two weeks of age onwards they developed a whole range of conditions that mimicked the ageing process. These included loss of fat under the skin, spinal curvature, and metabolism deficits. The mice died by one month of age, whereas a normal mouse can live for up to two years under laboratory conditions.

Most histone deacetylases are very promiscuous. By this we mean they will deacetylate any acetylated histone they can find. Indeed, as mentioned above, many don't even restrict themselves to histones, and will take acetyl groups off all sorts of proteins. However, SIRT6 isn't like this. It only takes the acetyl groups off two specific amino acids – lysine 9 and lysine 56, both on histone H3. The enzyme also seems to have a preference for histones that are positioned at telomeres. When Katrin Chua knocked out the *SIRT6* gene in human cells, she found that the telomeres of these cells got damaged, and the chromosomes began to join up. The cells lost the ability to divide any further and pretty much shut down most of their activities[13].

This suggested that human cells need SIRT6 so that they can maintain the healthy structures of telomeres. But this wasn't the only role of the SIRT6 protein. Acetylation of histone 3 at amino acid 9 is associated with gene expression. When SIRT6 removes this modification, this amino acid can be methylated by other enzymes present in the cell. Methylation at this position on the histone is associated with gene repression. Katrin Chua performed further experiments which confirmed that changing the expression levels of SIRT6 changed the expression of specific genes.

SIRT6 is targeted to specific genes by forming a complex with a particular protein. Once it's present at those genes, SIRT6 takes part in a feedback loop that keeps driving down expression of the gene, in a classic vicious cycle. When the *SIRT6* gene is knocked out, the levels of histone acetylation at these genes stays high because the feedback loop can't be switched on. This drives up expression of these target genes in the SIRT6 knockout mice. The target genes are ones which promote auto-destruction, or the cell's entry into a state of permanent stasis known as senescence. This effect explains why *SIRT6* knockdown is associated with premature ageing[14]. It's because genes that accelerate processes associated with ageing are switched on too soon, or too vigorously, at a young age.

It's a little like a crafty manufacturer installing an inbuilt obsolescence mechanism into a product. Normally, the mechanism doesn't kick in for a certain number of years, because if the obsolescence activates too early, the manufacturer will get a reputation for prematurely shoddy goods and nobody will buy them at all. Knocking out *SIRT6* in cells is a little like a software glitch that activates the inbuilt obsolescence pathway after, say, one month instead of two years.

Other *SIRT6* target genes are associated with provoking inflammatory and immune responses. This is also relevant to ageing, because some conditions that become much more common as we age are a result of increased activation of these pathways. These include certain aspects of cardiovascular disease and chronic conditions such as rheumatoid arthritis.

There is a rare genetic disease called Werner's syndrome. Patients with this disorder age faster and at an earlier age than healthy individuals. The condition is caused by mutations in a gene that is involved in the three-dimensional structure of DNA, keeping it in the correct conformation and wound up to the right degree of tightness for a specific cell type[15]. The normal protein binds to telomeres. It binds most effectively when the histones at the telomeres have lost the acetyl group at amino acid 9 on histone H3. This is the precise modification removed by the SIRT6 enzyme. This further strengthens the case for a role of SIRT6 in control of ageing[16].

Given that SIRT6 is a histone deacetylase, it might be interesting to test the effect of a histone deacetylase inhibitor on ageing. We would predict that it would have the same effects as knocking down expression of the SIRT6 enzyme, i.e. it would accelerate ageing. This might give us pause for thought when we plan to treat patients with histone deacetylase inhibitors such as SAHA. After all, an anti-cancer drug that makes you age faster isn't that attractive an idea.

Fortunately, from the point of view of treating cancer patients, SIRT6 belongs to a special class of histone deacetylase enzymes called sirtuins. Unlike the enzymes we met in Chapter 11, the sirtuins aren't affected by SAHA or any of the other histone deacetylase inhibitor drugs.

Eat less, live longer

All of this begs the question of whether we are any closer to finding a pill we can offer to people to increase longevity. The data so far don't seem promising, especially if it's true that many of the mechanisms that underlie ageing are defences against developing cancer. There's not a lot of point creating therapies that could allow us to live for another 50 years, if they also lead to tumours that could kill us in five. But there is one way of increasing lifespan that has proven astonishingly effective, from yeast to fruit flies, from worms to mammals. This is calorie restriction.

If you only give rodents about 60 per cent of the calories they would eat if given free access to food, there is a dramatic impact on longevity and development of age-related diseases[17]. The restricted calorie intake must start early in life and be continued throughout life to see this effect. In yeast, decreasing the amount of glucose (fuel) in the culture from 2 per cent to 0.5 per cent extended the lifespan by around 30 per cent[18].

There's been a lot of debate on whether or not this calorie-restriction effect is mediated via sirtuins, such as Sir2 in yeast, or the versions of Sir2 in other animals. Sir2 is regulated in part by a key chemical, whose levels are affected by the amount of nutrition available to cells. That's the reason why some authors have suggested that the two might be connected, and it's an attractive hypothesis. There's no debate that Sir2 is definitely important for longevity. Calorie restriction is also clearly very important. The question is whether the two work together or separately. There's no consensus as yet on this, and the experimental findings are very influenced by the model system used. This can come down to details that at first glance might almost seem trivial, such as which strain of brewer's yeast is used, or exactly how much glucose is in the culture liquid.

The question of how calorie restriction works might seem much less important than the fact that it does. But the mechanism matters enormously if we're looking for an anti-ageing strategy, because calorie restriction has severe limitations for humans. Food has enormous social and cultural aspects, it's rarely just fuel for us. In addition to these psychological and sociological issues, calorie restriction has side effects. The most obvious ones are muscle wasting and loss of libido. It's not much of a surprise that when offered the chances of living longer, but with these side-effects, the majority of people find the prospect unattractive[19].

That's one of the reasons that a 2006 paper in *Nature*, led by David Sinclair at Harvard Medical School, created such a furore. The scientists studied the effects of a compound called resveratrol on health and survival in mice. Resveratrol is a complex compound synthesised by plants, including grapes. It is a constituent of red wine. At the time of the paper, resveratrol had already been shown to extend lifespan in yeast, *C. elegans* and fruit flies[20,21].

Professor Sinclair and his colleagues raised mice on very high calorie diets, and treated the mice with resveratrol for six months. At the end of this six-month period, they examined all sorts of health outcomes in the mice. All the mice which had been on the high calorie diets were fat, regardless of whether or not they had been treated with resveratrol. But the mice treated with resveratrol were healthier than the untreated fat mice. Their livers were less fatty, their motor skills were better, they had fewer diabetes symptoms. By the age of 114 weeks, the resveratrol-treated mice had a 31 per cent lower death rate than the untreated animals fed the same diet[22].

We can see immediately why this paper garnered so much attention. If the same effects could be achieved in humans, resveratrol would be a get-out-of-obesity-free card. Eat as much as you like, get as fat as you want and yet still have a long and healthy life. No leaving behind one-third of every meal and losing your muscles and your libido.

How was resveratrol doing this? A previous paper from the same group showed that resveratrol activated a sirtuin protein, in this case Sirt1[23]. Sirt1 is believed to be important for the control of sugar and fat metabolism.

Professor Sinclair set up a company called Sirtris Pharmaceuticals, which continued to make new compounds based around the structure of resveratrol. In 2008 GlaxoSmithKline paid $720 million for Sirtris Pharmaceuticals to gain access to its expertise and portfolio of compounds for treating diseases of ageing.

This deal was considered expensive by many industry observers, and it hasn't been without its problems. In 2009, a group from rival pharmaceutical company Amgen published a paper. They claimed that resveratrol did not activate Sirt1, and that the original findings represented an artefact caused by technical problems[24]. Shortly afterwards, scientists from Pfizer, another pharmaceutical giant, published very similar findings to Amgen[25].

It's actually very unusual for large pharmaceutical companies to publish work that simply contradicts another company's findings. There's nothing much to be gained by doing so. Pharmaceutical companies are ultimately judged by the drugs they manage to launch successfully, and criticising a competitor in the early stages of a drug discovery programme gives them no commercial advantage. The fact that both Amgen and Pfizer went public with their findings is a demonstration of how controversial the resveratrol story had become.

Does it matter how resveratrol works? Isn't the most important feature the fact that it has such dramatic effects? If you are trying to develop new drugs to treat human conditions, it unfortunately matters quite a lot. The authorities who license new drugs are much keener on compounds when they know how they work. This is partly because this makes it much easier to monitor for side-effects, as you can develop better theories about what to look out for. But the other issue is that resveratrol itself probably isn't the ideal compound to use as a drug.

This is often an issue with natural products such as resveratrol, which was isolated from plants. The natural compounds may need to be altered to a greater or lesser extent, so that they circulate well in the body, and don't have unwanted side effects. For example, artemisinin is a chemical derived from wormwood which can kill malarial parasites. Artemisinin itself isn't taken up well by the human body so researchers developed compounds that were variants of the chemical structure of the original natural product. These variants kill malarial parasites, but are also much better than artemisinin at getting taken up by our bodies[26].

But if we don't know exactly how a particular compound is working, it's very hard to design and test new ones, because we don't know how to easily check if the new compounds are still affecting the right protein.

GlaxoSmithKline is standing by its sirtuin programmes, but in a worrying development for the company they have stopped a clinical trial of a special formulation of resveratrol in a disease called multiple myeloma, because of problems with kidney toxicity[27].

The progress of sirtuin histone deacetylase activators is of keen interest to all the big players in the pharmaceutical industry. We don't know yet if these epigenetic modifiers will set the agenda, or sound the death knell, for development of therapies specifically aimed at increasing longevity or combatting old age. So, for now, we're still stuck with our old routine: lots of vegetables, plenty of exercise and try to avoid harsh overhead lighting – it does nobody any favours.

Chapter 14

Long Live the Queen

All my possessions for a moment of time.
Attributed to Queen Elizabeth I

The effects of nutrition on the health and lifespan of mammals are pretty dramatic. As we saw in the previous chapter, prolonged calorie restriction can extend lifespan by as much as one-third in mice[1]. We also saw in Chapter 6 that our own health and longevity can be affected by the ways our parents and grandparents ate. These are quite startling findings but nature has provided us with a much more dramatic example of the impact of nutrition on lifespan. Imagine, if you can, a dietary regime that means a select few in a species have a lifespan that is twenty times longer than that of most of their companions. Twenty times longer. If that happened in humans, the UK might still be in the reign of Queen Elizabeth I, and would expect to be so for about another 400 years.

Obviously this doesn't happen in humans, but it does happen in one common organism. It's a creature that we all meet every spring and summer. We use the results of its labour to make candles and furniture polish, and we have eaten its hard-earned bounty since the very beginning of human history. It's the honeybee.

The honeybee, *Apis mellifera*, is a truly extraordinary creature. It is a prime example of a social insect. It lives in colonies that can contain tens of thousands of individuals. The vast majority of these are workers. These are sterile females, which have a range of specialised roles including gathering pollen, building living quarters and looking after the young. There are a small number of males, who do very little except mate, if they are lucky. And there is a queen.

In the formation of a new colony, a virgin queen leaves a hive, accompanied by a swarm of workers. She'll mate with some males and then settle down to form a new colony. The queen will lay thousands of eggs, most of which will hatch and develop into more workers. A few eggs will hatch and develop into new queens, who can start the whole cycle all over again.

Because the queen who founded the colony mated several times, not all the bees in the colony will be genetically identical to each other, because some of them will have different fathers. But there will be groups of thousands and thousands of genetically identical bees in any colony. This genetic identity doesn't refer only to the worker bees. The new queens are genetically identical to thousands of worker bees in the colony. We could call them sisters, but this doesn't really describe them well enough. They are all clones.

However, a new queen and her clonal worker sisters are clearly incredibly different from each other, both in physical form and in activities. The queen can be twice the size of a worker bee. After the so-called nuptial flight, when she first leaves a colony and mates, the queen almost never leaves the hive again. She stays in the darkness of the interior, laying up to 2,000 eggs a day in the summer months. She has no sting barbs, no wax glands and no pollen baskets (not much point having a shopping bag if you never leave the house). Worker bees have a lifespan that can usually be measured in weeks, whereas queens live for years[2].

Conversely, workers can do many things that the queens can't. Chief amongst these is collecting food, and then telling the rest of the colony its location. This information is communicated using the famous 'waggle dance'. The queen lives in darkened luxury, but she never gets to boogie.

So, a honeybee colony contains thousands of individuals who are genetically identical, but a few of them are really different physically and behaviourally. This difference in outcome is all down to how the bee larvae are fed. The pattern of early feeding completely determines whether a larva will develop into a worker or into a queen.

For honeybees the DNA script is constant but the outcome is variable. The outcome is controlled by an early event (feeding pattern) which sets a phenotype that is maintained throughout the rest of life. This is a scenario that just shrieks epigenetics at us, and in the last few years scientists have started to unravel the molecular events that underpin this process.

The critical roll of the dice for honeybees happens after the third day of life, as a fairly immobile grub or larva. Until day three, all honeybee larvae are given the same food. This is a substance called royal jelly, which is produced by a specialised group of workers. These young workers are known as nurse bees and they secrete royal jelly from glands in their heads. Royal jelly is a highly nutritious food source. It is a concentrated mix of a lot of different components, including key amino acids, unusual fats, specific proteins, vitamins and other nutrients that haven't been well-characterised yet.

Once the larvae are three days old, the nurse bees stop feeding royal jelly to most of them. Instead, most larvae are switched onto a diet of pollen and nectar. These are the larvae which will grow up to be worker bees.

But for reasons that nobody really understands, the nurse bees continue feeding royal jelly to a select few larvae. We don't know how these larvae are chosen or why. Genetically they are identical to the ones that are switched onto the less sophisticated diet. But this small group of larvae that continue to be nourished with royal jelly grow up to be queens, and they're fed this same substance throughout their lives. The royal jelly is essential for the production of mature ovaries in the queens. Worker females never develop proper ovaries, which is one of the reasons they are infertile. Royal jelly also prevents the queen from developing the organs that she won't ever need, like those pollen baskets.

We understand some of the mechanisms behind this process. Bee larvae contain an organ that has some of the same functions as our liver. When a larva receives royal jelly continuously, this organ processes the complex food source and activates the insulin pathway. This is very similar to the hormonal pathway in mammals that controls the levels of sugar in the bloodstream. In honeybees activation of this pathway increases production of another hormone, called Juvenile Hormone. Juvenile Hormone in turn activates other pathways. Some of these stimulate growth and development of tissues like the maturing ovaries. Others shut down production of the organs that the queen doesn't need[3].

Mimicking royalty

Because honeybee maturation has so many hallmarks of an epigenetic phenomenon, researchers speculated that there would also be an involvement of the epigenetic machinery. The first indications that this is indeed the case came in 2006. This was the year when researchers sequenced the genome of this species, to identify the fundamental genetic blueprint[4]. Their research showed that the honeybee genome contained genes that looked very similar to the DNA methyltransferase genes of higher organisms such as vertebrates. The honeybee genome was also shown to contain a lot of CpG motifs. This is the two-nucleotide sequence that is usually the target for DNA methyltransferases.

In the same year, a group led by Gene Robinson in Illinois showed that the predicted DNA methyltransferase proteins encoded in the honeybee genome were active. The proteins were able to add methyl groups to the cytosine residue in a CpG motif in DNA[5]. Honeybees also expressed proteins that were able to bind to methylated DNA. Together, these data showed that honeybee cells could both 'write' and 'read' an epigenetic code.

Until these data were published, nobody had really wanted to take a guess as to whether or not honeybees would possess a DNA methylation system. This was because the most widely used experimental system in insects, the fruit fly *Drosophila melanogaster*, whom we met earlier in this book, doesn't methylate its DNA.

It's interesting to discover that honeybees have an intact DNA methylation system. But this doesn't prove that DNA methylation is involved in the responses to royal jelly, or the persistent effects of this foodstuff on the physical form and behaviour of mature bees. This issue was addressed by some elegant work from the laboratory of Dr Ryszard Maleszka at the Australian National University in Canberra.

Dr Maleszka and his colleagues knocked down the expression of one of the DNA methyltransferases in honeybee larvae, by switching off the *Dnmt3* gene. Dnmt3 is responsible for adding methyl groups to regions of DNA that haven't been methylated before. The results of this experiment are shown in Figure 14.1.

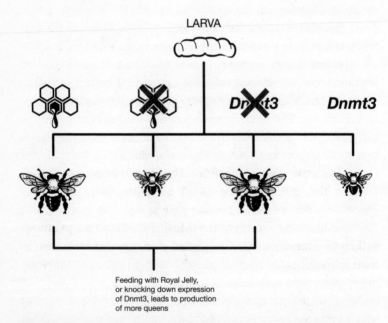

Figure 14.1   When royal jelly is fed to honeybee larvae for extended periods, the larvae develop into queens. The same effect is seen in the absence of prolonged feeding with royal jelly if the expression of the *Dnmt3* gene is decreased experimentally in the larvae. Dnmt3 protein adds methyl groups to DNA.

When the scientists decreased the expression of *Dnmt3* in the honeybee larvae, the results were the same as if they had fed them royal jelly. Most of the larvae matured as queens, rather than as workers. Because knocking down *Dnmt3* had the same effects as feeding royal jelly, this suggested that one of the major effects of royal jelly is connected with alterations of the DNA methylation patterns on important genes[6].

To back up this hypothesis, the researchers also examined the actual DNA methylation and gene expression patterns in the different experimental groups of bees. They showed that the brains of queens and worker bees have a different DNA methylation pattern. The DNA methylation patterns in the bees where *Dnmt3* had been knocked down were like those of the normal royal jelly-induced queens. This is what we would expect given the similar phenotypes in the two groups. The gene expression patterns in the normal queens and the *Dnmt3*-knockdown queens were also very similar. The authors concluded that the nutritional effects of continual feeding on royal jelly occurred via DNA methylation.

There are still a lot of gaps in our understanding of how nutrition in the honeybee larva results in altered patterns of DNA methylation. One hypothesis, based on the experiments above, is that royal jelly inhibits the DNA methyltransferase enzyme. But so far nobody has been able to demonstrate this experimentally. It's therefore possible that the effect of royal jelly on DNA methylation is indirect.

What we do know is that royal jelly influences hormonal signalling in honeybees, and that this changes gene expression patterns. Changes in the levels of expression of a gene often influence the epigenetic modifications at that gene. The more highly a gene is switched on, the more its histones become modified in ways which promote gene expression. Something similar may happen in honeybees.

We also know that the DNA methylation systems and histone modification systems often work together. This has created interest in the role of histone-modifying enzymes in the control of honeybee development and activity. When the honeybee genome was sequenced, four histone deacetylase enzymes were identified. This was intriguing because it has been known for some time that royal jelly contains a compound called phenyl butyrate[7]. This very small molecule can inhibit histone deacetylases but it does so rather weakly. In 2011, a group led by Dr Mark Bedford from the MD Anderson Cancer Center in Houston published an intriguing study on another component of royal jelly. One of the other senior authors on this paper was Professor Jean-Pierre Issa, who has been so influential in promoting use of epigenetic drugs for the treatment of cancer.

The researchers analysed a compound found in royal jelly called (E)-10-hydroxy-2-decenoic acid, or 10HDA for short. The structure of this compound is shown in Figure 14.2, along with SAHA, the histone deacetylase inhibitor we saw in Chapter 11 that is licensed for the treatment of cancer.

The two structures aren't identical by any means, but they do share some similarities. Each has a long chain of carbon atoms (the bit that looks like a crocodile's back in profile), and the right hand side of each compound also looks fairly similar. Mark Bedford and his colleagues hypothesised that 10HDA might be

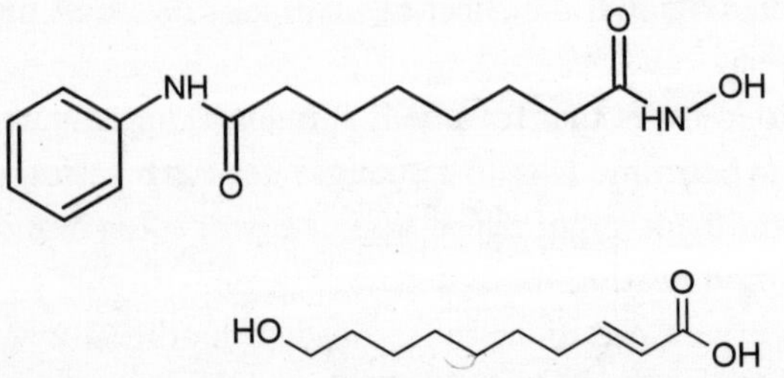

Figure 14.2   The chemical structure of the histone deacetylase inhibitor SAHA and 10HDA, a compound found in royal jelly. C: carbon; H: hydrogen; N: nitrogen; O: oxygen. For simplicity, some carbon atoms have not been explicitly shown, but are present where there is a junction of two lines.

an inhibitor of histone deacetylases. They performed a number of test tube and cell experiments and showed that this was indeed the case. This means that we now know that one of the major compounds found in royal jelly inhibits a key class of epigenetic enzymes[8].

The forgetful bee and the flexible toolkit

Epigenetics influences more than whether bees develop into workers or queens. Ryszard Maleszka has also shown that DNA methylation is involved in how honeybees process memories. When honeybees find a good source of pollen or nectar, they fly back to the hive and tell the other members of the colony where to head to find this great food supply. This tells us something really important about honeybees; they can remember information. They must be able to, or they wouldn't be able to tell the other bees where to go for food. Of course, it's equally important that the bees can forget information and replace it with new data. There's no point sending your co-workers to the great patch of thistles that were in flower last week, but that have now been eaten by a grazing donkey. The bee needs to forget last week's thistles and remember the location of this week's lavender.

It's actually possible to train honeybees to respond to certain stimuli associated with food. Dr Maleszka and his colleagues showed that when the bees undergo this training, the levels of Dnmt3 protein go up in the parts of the honeybee brains which are important in learning. If the bees are treated with a drug that inhibits the Dnmt3 protein, the compound alters the way the bees retain memories, and also the speed with which memories are lost[9].

Although we know that DNA methylation is important in honeybee memory, we don't know exactly how this works. This is because it's not clear yet which genes become methylated when honeybees learn and acquire new memories.

So far, we could be forgiven for thinking that honeybees and higher organisms, including us and our mammalian relatives, all use DNA methylation in the same way. It's certainly true that changes in DNA methylation are associated with alterations in developmental processes in both humans and honeybees. It's also true that mammals and honeybees both use DNA methylation in the brain during memory processing.

But oddly enough, honeybees and mammals use DNA methylation in very different ways. A carpenter has a saw in his toolbox and uses it to build a book case. An orthopaedic surgeon has a saw on his operating trolley and uses it to amputate a leg. Sometimes, the same bit of kit can be used in very different ways. Mammals and honeybees both use DNA methylation as a tool, but during the course of evolution they've employed it very differently.

When mammals methylate DNA, they usually methylate the promoter regions of genes, and not the parts that code for amino acids. Mammals also methylate repetitive DNA elements and transposons, as we saw in Emma Whitelaw's work in Chapter 5. DNA methylation in mammals tends to be associated with switching off gene expression and shutting down dangerous elements like transposons that might otherwise cause problems in our genomes.

Honeybees use DNA methylation in a completely different way. They don't methylate repetitive regions or transposons, so they presumably have other ways of controlling these potentially troublesome elements. They methylate CpG motifs in the stretches of genes that encode amino acids, rather than in the promoter regions of genes. Honeybees don't use DNA methylation to switch off genes. In honeybees, DNA methylation is found on genes that are expressed in all tissues, and also on genes that tend to be expressed by many different insect species. DNA methylation acts as a fine-tuning mechanism in honeybee tissues. It modulates the activity of genes, turning the volume slightly up or down, rather than acting as an on-off switch[10]. Patterns of DNA methylation are also strongly correlated with control of mRNA splicing in honeybee tissues. However, we don't yet know how this epigenetic modification actually influences the way in which a message is processed[11].

We're really only just beginning to unravel the subtleties of epigenetic regulation in honeybees. For example, there are 10,000,000 CpG sites in the honeybee genome, but less than 1 per cent of these are methylated in any given tissue. Unfortunately, this low degree of methylation makes analysing the effects of this epigenetic modification very challenging. The effects of *Dnmt3* knockdown show that DNA methylation is very important in honeybee development. But, given that DNA methylation is a fine-tuning mechanism in this species, it's likely that *Dnmt3* knockdown results in a number of individually minor changes in a relatively large number of genes, rather than dramatic changes in a few. These types of subtle alterations are the most difficult to analyse, and to investigate experimentally.

Honeybees aren't the only insect species that has developed a complex society with differing forms and functions for genetically identical individuals. This model has evolved independently several times, including in different species of wasps, termites, bees and ants. We don't yet know if the same epigenetic processes are used in all these cases. Shelley Berger from the University of Pennsylvania, whose work on ageing we encountered in Chapter 13, is involved in a large collaboration focusing on ant genetics and epigenetics. This work has already shown that at least two species of ants also can methylate the DNA in their genomes. The expression of different epigenetic enzymes varies between different social groups in the colonies[12]. These data tentatively suggest that epigenetic control of colony members may prove to be a mechanism that has evolved more than once in the social insects.

For now, however, most interest in the world outside epigenetics labs focuses on royal jelly, as this has a long history as a health supplement. It's worth pointing out that there's very little hard evidence to support this having any major effects in humans. The 10HDA, that Mark Bedford and his colleagues showed was a histone deacetylase inhibitor, can affect the growth of blood vessel cells[13]. Theoretically, this could be useful in cancer, as tumours rely on a good blood supply for continuing growth. However, we're a very long way from showing that royal jelly can really fight off cancer, or aid human health in any other way. If there's one thing we do already know, it's that bees and humans are not the same epigenetically. Which is just as well, unless you're a really big fan of the monarchy …

Chapter 15

The Green Revolution

To see a world in a Grain of Sand,
And a Heaven in a Wild Flower,
Hold Infinity in the palm of your hand,
And eternity in an hour.
William Blake

Probably all of us are familiar with the guessing game 'animal, vegetable or mineral'. The implicit assumption in the name of this game is that plants and animals are completely different from one another. True, they are both living organisms, but that's where we feel the similarity ends. We may be able to get on board with the idea that somewhere back in the murky evolutionary past, humans and microscopic worms have a shared ancestor. But how often do we ever wonder about the biological heritage we share with plants? When do we ever think of carnations as our cousins?

Yet animals and plants are surprisingly similar in many ways. This is especially the case when we consider the most advanced of our green relatives, the flowering plants. These include the grasses and cereals that we rely on for so much of our basic food intake, and the broad-leaved plants, from cabbages to oak trees and from rhododendrons to cress.

Animals and the flowering plants are each made up of lots of cells; they are multi-cellular organisms. Many of these cells are specialised for particular functions. In the flowering plants these include cells that transport water or sugars around the plant, the photosynthesising cells of the leaves and the food storing cells of the roots. Like animals, plants have specialised cells which are responsible for sexual reproduction. The sperm nuclei are carried in pollen and fertilise a large egg cell, which ultimately gives rise to a zygote and a new individual plant.

The similarities between plants and animals are more fundamental than these visible features. There are many genes in plants which have equivalents in animals. Crucially, for our topic, plants also have a highly developed epigenetic system. They can modify histone proteins and DNA, just like animal cells can, and in many cases use very similar epigenetic enzymes to those used by animals, including humans.

These genetic and epigenetic similarities all suggest that animals and plants have common ancestors. Because of our common ancestry, we've inherited similar genetic and epigenetic tool kits.

Of course, there are also really important differences between plants and animals. Plants can create their own food, but animals can't do this. Plants take in basic chemicals in the environment, especially water and carbon dioxide. Using energy from sunlight, plants can convert these simple chemicals into complex sugars such as glucose. Nearly all life on planet earth is dependent directly or indirectly on this amazing process of photosynthesis.

There are two other ways in which plants and animals are very different. Most gardeners know that you can take a cutting from a growing plant – maybe just a small shoot – and create an entire new plant from this. There are very few animals where this is possible, and certainly no advanced ones. True, if certain species of lizard lose their tail, the animal can grow a new one. But they can't do this the other way around. We can't grow a new lizard from a discarded bit of tail.

This is because in most adult animals the only genuinely pluripotent stem cells are the tightly controlled cells of the germline which give rise to eggs or sperm. But active pluripotent stem cells are a completely normal part of a plant. In plants these pluripotent stem cells are found at the tips of stems and the tips of roots. Under the right conditions, these stem cells can keep dividing to allow the plant to grow. But under other conditions, the stem cells will differentiate into specific cell types, such as flowers. Once such a cell has become committed to becoming part of a petal, for example, it can't change back into a stem cell. Even plant cells roll down Waddington's epigenetic landscape eventually.

The other difference between plants and animals is really obvious. Plants can't move. When environmental conditions change, the plant must adapt or die. They can't out-run or out-fly unfavourable climates. Plants have to find a way of responding to the environmental triggers all around them. They need to make sure they survive long enough to reproduce at the right time of year, when their offspring will have the greatest chance of making it as new individuals.

Contrast this with a species such as the European swallow (*Hirundo rustica*) which winters in South Africa. As summer approaches and conditions become unbearable the swallow sets off on an epic migration. It flies up through Africa and Europe, to spend the summer in the UK where it raises its young. Six months later, back it goes to South Africa.

Many of a plant's responses to the environment are linked to changes in cell fate. These include the change from being a pluripotent stem cell to becoming part of a terminally differentiated flower in order to allow sexual reproduction. Epigenetic processes play important roles in both these events, and interact with other pathways in plant cells to maximise the chance of reproductive success.

Not all plants use exactly the same epigenetic strategies. The best-characterised model system is an insignificant looking little flowering plant called *Arabidopsis thaliana*. It's a member of the mustard family and looks like any nondescript weed you can find on any patch of wasteland. Most of the leaves grow close to the ground in a rosette shape. It produces small white flowers on a stem about 20–25 centimetres high. It's been a useful model system for researchers because its genome is very compact, which makes it easy to sequence in order to identify the genes. There are also well-developed techniques for genetically modifying *Arabidopsis thaliana*. This makes it relatively straightforward for scientists to introduce mutations into genes to investigate their function.

Arabidopsis thaliana seeds typically germinate in early summer in the wild. The seedlings grow, creating the rosette of leaves. This is called the vegetative phase of plant growth. In order to produce offspring, *Arabidopsis thaliana* generates flowers. It is structures in the flowers that will generate the new eggs and sperm that will eventually lead to new zygotes, which will be dispersed in seeds.

But here's the problem for the plant. If it flowers late in the year, the seeds it produces will be wasted. That's because the weather conditions won't be right for the new seeds to germinate. Even if the seeds do manage to germinate, the tender little seedlings are likely to be killed off by harsh weather like frost.

The adult *Arabidopsis thaliana* needs to keep its powder dry. It has a much greater chance of lots of its offspring surviving if it waits until the next spring until it flowers. The adult plant can survive winter weather that would kill off a seedling. This is exactly what *Arabidopsis thaliana* does. The plant 'waits' for spring and only then does it produce flowers.

The rites of spring

The technical term for this is vernalisation. Vernalisation means that a plant has to undergo a prolonged cold period (winter, usually) before it can flower. This is very common in plants with an annual life-cycle, especially in the temperate regions of the earth where the seasons are well-defined. Vernalisation doesn't just affect broad-leaved plants like *Arabidopsis thaliana*. Many cereals also show this effect, especially crops like winter barley and winter wheat. In many cases, the prolonged period of cold needs to be followed by an increase in day length if flowering is to take place. The combination of the two stimuli ensures that flowering occurs at the most appropriate time of year.

Vernalisation has some very interesting features. When the plant first begins to sense and respond to cold weather, this may be many weeks or months before it starts to flower. The plant may continue to grow vegetatively through cell division during the cold period. When new seeds are produced, after the vernalisation of the parent plant, the seeds are 'reset'. The new plants they produce from the seeds will themselves have to go through their own cold season before flowering[1].

These features of vernalisation are all very reminiscent of epigenetic phenomena in animals. Specifically:

1. The plant displays some form of molecular memory, because the stimulus and the final event are separated by weeks or months. We can compare this with abnormal stress responses in adult rodents that were 'neglected' as infants.
2. The memory is maintained even after cells divide. We can compare this with animal cells that continue to perform in a certain way after a stimulus to the parent cell, such as in normal development or in cancer progression.
3. The memory is lost in the next generation (the seeds). This is comparable with the way that most changes to the somatic tissues are 'wiped clean' in animals so that Lamarckian inheritance is exceptional, rather than common.

So, at a phenomenon level, vernalisation looks very epigenetic. In recent years, a number of labs have confirmed that epigenetic processes underlie this, at the chromatin modification level.

The key gene involved in vernalisation is called *FLOWERING LOCUS C* or *FLC* for short. *FLC* encodes a protein called a transcriptional repressor. It binds to other genes and stops them getting switched on. There are three genes that are particularly important for flowering in *Arabidopsis thaliana*, called *FT*, *SOC1* and *FD*. Figure 15.1 shows how *FLC* interacts with these genes, and the consequences this has for flowering. It also shows how

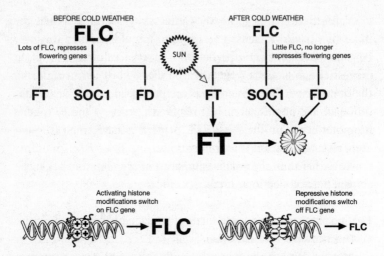

Figure 15.1 Epigenetic modifications regulate the expression of the *FLC* gene, which represses the genes which promote flowering. The epigenetic modifications on the *FLC* gene are controlled by temperature.

the epigenetic status of FLC changes after a period of prolonged cold.

Before winter, the *FLC* gene promoter carries lots of histone modifications that switch on gene expression. Because of this, the *FLC* gene is highly expressed, and the protein it codes for binds to the target genes and represses them. This keeps the plant in its normal growing vegetative phase. After winter, the histone modifications at the *FLC* gene promoter change to repressive ones. These switch off the *FLC* gene. The FLC protein levels drop, which removes the repression on the target genes. The increased periods of sunlight during spring activate expression of the *FT* gene. It's essential that FLC levels have gone down by this stage, because if FLC levels are high, the *FT* gene finds it difficult to react to the stimulus from sunlight[2].

Experiments with mutated versions of epigenetic enzymes have shown that the changes in histone modifications at the *FLC* gene are critically important in controlling the flowering response. For example, there is a gene called *SDG27* which adds methyl groups to the lysine amino acid at position 4 on histone H3, so it is an epigenetic writer. This methylation is associated with active gene expression. The *SDG27* gene can be mutated experimentally, so that it no longer encodes an active protein. Plants with this mutation have less of this active histone modification at the *FLC* gene promoter. They produce less FLC protein, and so aren't so good at repressing the genes that trigger flowering. The *SDG27* mutants flower earlier than the normal plants[3]. This demonstrates that the epigenetic modifications at the *FLC* promoter don't simply reflect the activity levels of the gene, they actually alter the expression. The modifications do actually cause the change in expression.

Cold weather induces a protein in plant cells called VIN3. This protein can bind to the *FLC* promoter. VIN3 is a type of protein called a chromatin remodeller. It can change how tightly chromatin is wound up. When VIN3 binds to the *FLC* promoter, it alters the local structure of the chromatin, making it more accessible to other proteins. Often, opening up chromatin leads to an increase in gene expression. However, in this case, VIN3 attracts yet another enzyme that can add methyl groups to histone proteins. However, this particular enzyme adds methyl groups to the lysine amino acid at position 27 on histone H3. This modification represses gene expression and is one of the most important methods that the plant cell uses to switch off the *FLC* gene[4,5].

This still raises the question of how cold weather results in epigenetic changes to the *FLC* gene specifically. What is the targeting mechanism? We still don't know all the details, but one of the stages has been elucidated. Following cold weather, the cells in *Arabidopsis thaliana* produce a long RNA, which doesn't code for protein. This RNA is called COLDAIR. The COLDAIR noncoding RNA is localised specifically at the *FLC* gene. When localised, it binds to the enzyme complex that creates the important repressive mark at position 27 on histone H3. COLDAIR therefore acts as a targeting mechanism for the enzyme complex[6].

When *Arabidopsis thaliana* produces new seeds, the repressive histone marks at the *FLC* gene are removed. They are replaced by activating chromatin modifications. This ensures that when the seeds germinate the *FLC* gene will be switched on, and repress flowering until the new plants have grown through winter.

From these data we can see that flowering plants clearly use some of the same epigenetic machinery as many animal cells. These include modifications of histone proteins, and the use of long non-coding RNAs to target these modifications. True, animal and plant cells use these tools for different end-points – remember the orthopaedic surgeon and the carpenter from the previous chapter – but this is strong evidence for common ancestry and one basic set of tools.

The epigenetic similarities between plants and animals don't end here either. Just like animals, plants also produce thousands of different small RNA molecules. These don't code for proteins, instead they silence genes. It was scientists working with plants who first realised that these very small RNA molecules can move from one cell to another, silencing gene expression as they go[7,8]. This spreads the epigenetic response to a stimulus from one initial location to distant parts of the organism.

The kamikaze cereal

Research in *Arabidopsis thaliana* has shown that plants use epigenetic modifications to regulate thousands of genes[9]. This regulation probably serves the same purposes as in animal cells. It helps cells to maintain appropriate but short-term responses to environmental stimuli, and it also locks differentiated cells in permanent patterns of specific gene expression. Because of epigenetic mechanisms we humans don't have teeth in our eyeballs, and plants don't have leaves growing out of their roots.

Flowering plants share a characteristic epigenetic phenomenon with mammals, and with no other members of the animal kingdom. Flowering plants are the only organisms we know of besides placental mammals in which genes are imprinted. Imprinting is the process we examined in Chapter 8, where the expression pattern of a gene is dependent on whether it was inherited from the mother or father.

At first glance, this similarity between flowering plants and mammals seems positively bizarre. But there's an interesting parallel between us and our floral relations. In all higher mammals, the fertilised zygote is the source of both the embryo and the placenta. The placenta nourishes the developing embryo, but doesn't ultimately form part of the new individual. Something rather similar happens when fertilisation occurs in flowering plants. The process is slightly more complicated, but the final fertilised seed contains the embryo and an accessory tissue called the endosperm, shown in Figure 15.2.

Just like the placenta in mammalian development, the endosperm nourishes the embryo. It promotes development and germination but it doesn't contribute genetically to the next

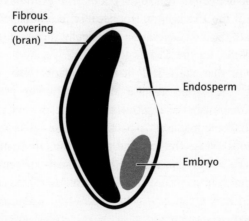

Figure 15.2 The major anatomical components of a seed. The relatively small embryo that will give rise to the new plant is nourished by the endosperm, in a manner somewhat analogous to the nourishment of mammalian embryos by the placenta.

generation. The presence of any accessory tissues during development, be this a placenta or an endosperm, seems to favour the generation of imprinted control of the expression of a select group of genes.

In fact, something very sophisticated happens in the endosperm of seeds. Just like most animal genomes, the genomes of flowering plants contain retrotransposons. These are usually referred to as TEs – transposable elements. These are the repetitive elements that don't encode proteins, but can cause havoc if they are activated. This is especially because they can move around in the genome and disrupt gene expression.

Normally such TEs are tightly repressed, but in the endosperm these sequences are switched on. The cells of the endosperm create small RNA molecules from these TEs. These small RNAs travel out from the endosperm into the embryo. They find the TEs in the embryo's genome that have the same sequence as themselves. These TE small RNA molecules then seem to recruit the machinery that permanently inactivates these potentially dangerous genomic elements. The risk to the endosperm genome through re-activation of the TEs is high. But because the endosperm doesn't contribute to the next generation genetically, it can undertake this suicide mission, for the greater good[10,11,12,13].

Although mammals and flowering plants both carry out imprinting, they seem to use slightly different mechanisms. Mammals inactivate the appropriate copy of the imprinted gene by using DNA methylation. In plants, the paternally-derived copy of the gene is always the one that carries the DNA methylation. However, it's not always this methylated copy of the gene that is inactivated[14]. In plant imprinting, therefore, DNA methylation tells the cell how a gene was inherited, not how the gene should be expressed.

There are some fundamental aspects of DNA methylation that are quite similar between plants and animals. Plant genomes encode active DNA methyltransferase enzymes, and also proteins that can 'read' methylated DNA. Just like primordial germ cells in mammals, certain plant cells can actively remove methylation from DNA. In plants, we even know which enzymes carry out this reaction[15]. One is called DEMETER, after the mother of Persephone in Greek myths. Demeter was the goddess of the harvest and it was because of the deal that she struck with Hades, the god of the Underworld, that we have seasons.

But DNA methylation is also an aspect of epigenetics where there are clear differences in the way plants and higher animals use the same basic system. One of the most obvious differences is that plants don't just methylate at CpG motifs (cytosine followed by a guanine). Although this is the most common sequence targeted by their DNA methyltransferases, plants will also methylate a cytosine followed by almost any other base[16].

A lot of DNA methylation in plants is focused around non-expressed repetitive elements, just like in mammals. But a big difference becomes apparent when we examine the pattern of DNA methylation in expressed genes. About 5 per cent of expressed plant genes have detectable DNA methylation at their promoters, but over 30 per cent are methylated in the regions that encode amino acids, in the so-called body of the genes. Genes with methylation in the body regions tend to be expressed in a wide range of tissues, and are expressed at moderate to high levels in these tissues[17].

The high levels of DNA methylation at repetitive elements in plants are very similar to the pattern at repetitive elements in the chromatin of higher animals such as mammals. By contrast, the methylation in the bodies of widely expressed genes is much more like that seen in honeybees (which don't methylate their repetitive elements). This doesn't mean that plants are some strange epigenetic hybrid of insects and mammals. Instead, it suggests that evolution has a limited set of raw materials, but isn't too obsessive about how it uses them.

Chapter 16

The Ways Ahead

Prediction is very difficult, especially about the future.
Niels Bohr

One of the most exciting things about epigenetics is the fact that in some ways it's very accessible to non-specialists. We can't all have access to the latest experimental techniques, so not all of us will unravel the chromatin changes that underlie epigenetic events. But all of us can examine the world around us and make predictions. All we need to do is look to see if a phenomenon meets the two most essential criteria in epigenetics. By doing this, we can view the natural world, including humans, in a completely new light. These two criteria are the ones we have returned to over and over again throughout this book. A phenomenon is likely to be influenced by epigenetic alterations in DNA and its accompanying proteins if one or both of the following conditions are met:

1. Two things are genetically identical, but phenotypically variable;
2. An organism continues to be influenced by an event long after this initiating event has occurred.

We always have to apply a common sense filter, of course. If someone loses their leg in a motorbike accident, the fact that they are still minus a leg twenty years later doesn't mean that we can invoke an epigenetic mechanism. On the other hand, that person may continue to have the sensation that they have both legs. This phantom limb syndrome might well be influenced by programmed gene expression patterns in the central nervous system that are maintained in part by epigenetic modifications.

We are sometimes so overwhelmed by the technologies used in modern biology that we forget how much we can learn just by looking thoughtfully. For example, we don't always need sophisticated laboratory equipment to determine if two phenotypically different things are genetically identical. Here are a couple of examples with which we are all familiar. Maggots turn into flies and caterpillars turn into butterflies. An individual maggot and the adult fly into which it finally develops must have the same genetic code. It's not as if a maggot can request a new genome as it metamorphoses. So, the maggot and the fly use the same genome in completely different ways. The painted lady caterpillar has interesting spikes all over its body and is fairly dull in colour. Like a maggot, it has no wings. The painted lady butterfly is a beautiful creature, with enormous wings coloured black and vivid orange, and it has no big spikes on its body. Once again, an individual caterpillar and the butterfly into which it develops must have the exact same DNA script. But the final productions from these scripts differ enormously. We can hypothesise that this is likely to involve epigenetic events.

The stoat *Mustela ermine* is found in Europe and North America. It's an athletic little predator in the weasel family, and in summer the fur on its back is a warm brown and its front is a creamy white. In cold climates its coat turns almost completely white all over in the winter, except for the tip of its tail, which remains black. With the arrival of spring, the stoat reverts to its summer colours. We know that there are hormonal effects that are required for this seasonal change in coat colour. It's pretty reasonable to hypothesise that these influence the relevant expression of coat colour genes by methods which include epigenetic modifications to chromatin.

In mammals, there's usually a clear genetic reason why males are males and females are females. A functional Y chromosome leads to the male phenotype. In lots of reptile species, including crocodiles and alligators, the two sexes are genetically identical.

You can't predict the sex of a crocodile from its chromosomes. The sex of a crocodile or an alligator depends on the temperature during critical stages in the development of the egg – the same blueprint can be used to create either a male or a female croc[1]. We know that hormonal signalling is involved in this process. There hasn't been much investigation of whether or not epigenetic modifications play a role in establishing or stabilising the gender-specific patterns of gene expression, but it seems likely.

Understanding the mechanisms of sex determination in crocodiles and their relatives may become a rather important conservation issue in the near future. The global shift in temperatures due to climate change could have adverse consequences for these reptiles, if the populations become very skewed in favour of one sex over another. Some authors have even speculated that such an effect may have contributed to the extinction of the dinosaurs[2].

The ideas above are quite straightforward, easily testable hypotheses. We can generate a lot more like these by simple observation. It's a lot riskier to make broad claims about what other more general developments we might expect to see in epigenetic research. The field is still young, and moving in all sorts of unexpected directions. But let's render ourselves hostages to fortune, and make a few predictions anyway.

We'll start with a fairly specific one. By 2016 at least one Nobel Prize for Physiology or Medicine will have been awarded to some leading workers in this field. The question is to whom, because there are plenty of worthy candidates.

For many people in the field it's extraordinary that this hasn't yet been awarded to Mary Lyon for her remarkably prescient work on X inactivation. Although her key papers that laid the conceptual framework for X inactivation didn't contain much original experimental data, this is also true of James Watson and Francis Crick's original paper on the structure of DNA[3]. It's always tempting to speculate the lack of a Nobel Prize might be down to gender, but that's partly because of a myth that has grown up around

Rosalind Franklin. She was the X-ray crystallographer whose data were essential for the development of the Watson-Crick model of DNA. When the Nobel Prize was awarded to Watson and Crick in 1962 it was also awarded to Rosalind Franklin's lab head, Professor Maurice Wilkins from Kings College, London. But Rosalind Franklin didn't miss out on the prize because she was a woman. She missed out because she had, tragically, died of ovarian cancer at the age of 37, and the Nobel Prize is never awarded posthumously.

Bruce Cattanach is a scientist we have met before in these pages. In addition to his work on parent-of-origin effects, he also performed some of the key early experimental studies on the molecular mechanisms behind X inactivation[4]. He would be considered a worthy co-recipient with Mary Lyon by most researchers. Mary Lyon and Bruce Cattanach performed much of their seminal research in the 1960s and are long-since retired. However, Robert Edwards, the pioneer of in vitro fertilisation, received the 2010 Nobel Prize in his mid-eighties, so there is still time and a little hope left for Professors Lyon and Cattanach.

The work of John Gurdon and Shinya Yamanaka on cellular reprogramming has revolutionised our understanding of how cell fate is controlled, and they must be hot favourites for a trip to Stockholm soon. A slightly less mainstream but appealing combination would be Azim Surani and Emma Whitelaw. Together their work has been seminal in demonstrating not only how the epigenome is usually reset in sexual reproduction, but also how this process is occasionally subverted to allow the inheritance of acquired characteristics. David Allis has led the field in the study of epigenetic modifications to histones, and must also be an attractive choice, possibly in combination with some of the leading lights in DNA methylation, especially Adrian Bird and Peter Jones.

Peter Jones has been a pioneer in the development of epigenetic therapies and this is another growth area for epigenetics. Histone deacetylase inhibitors and DNA methyltransferase inhibitors have been in the vanguard of these approaches. The vast majority of clinical trials with these compounds have been in cancer, but this is starting to change. An inhibitor of the sirtuin class of histone deacetylases is in early clinical trials for Huntington's disease, the devastating inherited neurodegenerative disorder[5]. The greatest excitement, for both cancer and non-oncology conditions, is currently centred around the development of drugs that inhibit more focused epigenetic enzymes. These include enzymes that change just one modification at one specific amino acid position on histone proteins. Hundreds of millions of dollars are being invested worldwide in this sphere, either in new biotech companies, or by the pharmaceutical giants. We are likely to see new drugs from these efforts enter clinical trials for cancer in the next five years, and clinical trials for other less immediately life-threatening conditions within a decade[6].

Our increased understanding of epigenetics, and especially of transgenerational inheritance, may also create problems in drug discovery, as well as opportunities. If we create new drugs that interfere with epigenetic processes, what if these drugs also affect the reprogramming that normally occurs during the production of germ cells? This could theoretically result in physiological changes that don't just affect the person who was treated, but also their children or grandchildren. We maybe shouldn't even restrict our concerns to chemicals that specifically target epigenetic enzymes. As we saw in Chapter 8, the environmental pollutant vinclozolin can affect rodents for many generations. If the authorities that regulate the licensing of new drugs begin to insist on transgenerational studies, this will add enormously to the cost and complexity of developing new drugs.

At first glance, this might seem perfectly reasonable; after all, we want drugs to be as safe as possible. But what happens to all the patients who desperately need new drugs to save them from life-threatening diseases, or who need better drugs so that they can live healthy and dignified lives free of pain and disability? The longer it takes to get new drugs to the market, the longer those patients suffer. It's going to be very interesting to see how drug companies, regulators and patient advocacy groups deal with this issue over the next ten or fifteen years.

Transgenerational effects of epigenetic changes may be one of the areas with the greatest impact on human health over the coming decades, not because of drugs or pollutants but because of food and nutrition. We started this journey into the epigenetic landscape by looking at the Dutch Hunger Winter. This had consequences not just for those who lived through it but for their descendants. We are in the grip of a global obesity epidemic. Even if our societies manage to get control of this (and very few western cultures show many signs of doing so) we may already have generated a less than optimal epigenetic legacy for our children and grandchildren.

Nutrition in general is one area where we can predict epigenetics will come to the fore in the next ten years. Here are just a few examples of what we know at the moment.

Folic acid is one of the supplements recommended for pregnant women. Increasing the supply of folic acid in the very early stages of pregnancy has been a public health triumph, as it has led to a major drop in the incidence of spina bifida in newborns[7]. Folic acid is required for the production of a chemical called SAM (S-adenosyl methionine). SAM is the molecule that donates the methyl group when DNA methyltransferases modify DNA. If baby rats are fed a diet that is low in folic acid, they develop abnormal regulation of imprinted regions of the genome[8]. We are only just beginning to unravel how many of the beneficial effects of folic acid may be mediated through epigenetic mechanisms.

Histone deacetylase inhibitors in our diets may also play useful roles in preventing cancer and possibly other disorders. The data are relatively speculative at the moment. Sodium butyrate in cheese, sulphoraphane in broccoli and diallyl disulphide in garlic are all weak inhibitors of histone deacetylases. Researchers have hypothesised that the release of these compounds from food during digestion may help to modulate gene expression and cell proliferation in the gut[9]. In theory, this could lower the risk of developing cancerous changes in the colon. The bacteria in our intestines also naturally produce butyrate from the breakdown of foodstuffs[10], especially plant-derived materials, which is another good reason to eat our greens.

There's a speculative but fascinating case study from Iceland on how diet may epigenetically influence a disease. It concerns a rare genetic disease called hereditary cystatin C amyloid angiopathy, which causes premature death through strokes. In the Icelandic families in which some people suffer from the disease, the patients carry a particular mutation in the key gene. Because of the relatively isolated nature of Icelandic societies, and the country's excellent record keeping, researchers were able to trace this disease back through the affected families. What they found was quite remarkable. Until about 1820, people with this mutation lived until around the age of 60 before they succumbed to the disease. Between 1820 and 1900, the life expectancy for those with the same disorder dropped to about 30 years of age, which is where it has remained. The scientists speculated in their original paper that an environmental change in the period from 1820 onwards altered the way that cells respond to and control the effects of the mutation[11].

At a conference in Cambridge in 2010, the same authors reported that one of the major environmental changes in Iceland from 1820 to the present day was a shift from a traditional diet to more mainstream European fare[12]. The traditional Icelandic diet contained exceptionally high quantities of dried fish and fermented butter. The latter is very high in butyric acid, the weak histone deacetylase inhibitor. Histone deacetylase inhibitors can alter the function of muscle fibres in blood vessels[13], which is relevant to the type of stroke that patients with this mutation suffer.

There is no formal proof yet that it's the drop in consumption of dietary histone deacetylase inhibitors that has led to the earlier deaths in this patient group, but it's a fascinating hypothesis.

The fundamental science of epigenetics is the area that is most difficult to make predictions about. One fairly safe bet is that epigenetic mechanisms will continue to crop up in unexpected parts of science. A good recent example of this is in the field of circadian rhythms, the natural 24-hour cycle of physiology and biochemistry found in most living species. A histone acetyltransferase has been shown to be the key protein involved in setting this rhythm[14], and the rhythm is adjusted by at least one other epigenetic enzyme[15].

We are also likely to find that some epigenetic enzymes influence cells in many different ways. That's because quite a few of these enzymes don't just modify chromatin. They can also modify other proteins in the cell, so may act on lots of different pathways at once. In fact, it has been proposed that some of the histone modifying genes actually evolved before cells contained histones[16]. This would suggest that these enzymes originally had other functions, and have been press-ganged by evolution into becoming controllers of gene expression. It wouldn't therefore be surprising to find that some of the enzymes have dual functions in our cells.

Some of the most fundamental issues around the molecular machinery of epigenetics remain very mysterious. Our knowledge of how specific modifications are established at selected positions in the genome is really sketchy. We are starting to see a role for non-coding RNAs in this process, but there are still multiple gaps in our understanding. Similarly we have almost no idea of how histone modifications are transmitted from mother cell to daughter cell. We're pretty sure this happens, as it is part of the molecular memory of cells that allows them to maintain cell fate, but we don't know how. When DNA is replicated, the histone proteins get pushed to one side. The new copy of the DNA may end up with relatively few of the modified histones. Instead, it may be coated with virgin histones with hardly any modifications. This is corrected very quickly, but we have almost no understanding of how this happens, even though it is one of the most fundamental issues in the whole field of epigenetics.

It's possible that we won't be able to solve this mystery until we have the technology and imagination to stop thinking in two dimensions and move to a three-dimensional world. We have become very used to thinking of the genome in linear terms, as strings of bases that are just read in a straightforward fashion. Yet the reality is that different regions of the genome bend and fold, reaching out to each other to create new combinations and regulatory sub-groups. We think of our genetic material as a normal script, but it's more like the fold-in from the back of *Mad* magazine, where folding an image in a particular way created a new picture. Understanding this process may be critical for truly unravelling how epigenetic modifications and gene combinations work together to create the miracle of the worm or the oak or the crocodile.

Or us.

So here's the summary of what epigenetic research will hold in the next decade. There will be hope and hype, over-promising, blind alleys, wrong turns and occasionally even some discredited research. Science is a human endeavour and sometimes it goes wrong. But at the end of the next ten years we will understand more of the answers to some of biology's most important questions. Right now we really can't predict what those answers might be, and in some cases we're not even sure of the questions, but one thing is for sure.

The epigenetics revolution is underway.

Notes

Introduction

1. For these and many others, see http://news.bbc.co.uk/1/hi/sci/tech/807126.stm

2. A useful starting point for descriptions of the symptoms of schizophrenia, its effects on patients and their families, and relevant statistics is www.schizophrenia.com

Chapter 1

1. http://www.britishlivertrust.org.uk/home/the-liver.aspx

2. http://www.wellcome.ac.uk/News/2010/Features/WTX063605.htm

3. Quoted in the *The Scientist Speculates*, ed. Good, I.J. (1962), published by Heinemann.

4. Key papers from this programme of work include: Gurdon et al. (1958) *Nature* 182: 64–5; Gurdon (1960) *J Embryol Exp Morphol.* 8: 505–26; Gurdon (1962) *J Hered.* 53: 5–9; Gurdon (1962) *Dev Biol.* 4: 256–73; Gurdon (1962) *J Embryol Exp Morphol.* 10: 622–40.

5. Waddington, C. H. (1957), *The Strategy of the Genes*, published by Geo Allen & Unwin.

6. Campbell et al. 1996 *Nature* 380: 64–6.

Chapter 2

1. For a useful review of the state of knowledge at the time see Rao, M. (2004) *Dev Biol.* 275: 269–86.

2. Takahashi and Yamanaka (2006), *Cell* 126: 663–76.

3. Pang et al. (2011), *Nature* online publication May 26.

4. Alipio et al. (2010), *Proc Natl Acad Sci. USA* 107: 13426–31.

5. Nakagawa et al. (2008), *Nat Biotechnol.* 26: 101–6.

6. See, for example, Baharvand et al. (2010) *Methods Mol Biol.* 584: 425–43.

7. Gaspar and Thrasher (2005), *Expert Opin Biol Ther.* 5: 1175–82.

8. Lapillonne et al. (2010), *Haematologica* 95: 1651–9.

Chapter 3

1. See http://genome.wellcome.ac.uk/doc_WTD020745.html for a wealth of useful genome-related facts and figures.
2. Schoenfelder et al. (2010), *Nat Genet.* 42: 53–61.

Chapter 4

1. Kruczek and Doerfler (1982), *EMBO J.* 1:409–14.
2. Bird et al. (1985), *Cell* 40: 91–99.
3. Lewis et al. (1992), *Cell* 69: 905–14.
4. Nan et al. (1998), *Nature* 393: 386–9.
5. For a recent review of the actions of MeCP2, see Adkins and Georgel (2011), *Biochem Cell Biol.* 89: 1–11.
6. Guy et al. (2007), *Science* 315: 1143–7.
7. http://www.youtube.com/watch?v=RyAvKGmAElQ&feature=related
8. The most important papers from the Allis lab in 1996 were: Brownell et al. (1996), *Cell* 84: 843–51; Vettese-Dadey et al. (1996), *EMBO J.* 15: 2508–18; Kuo et al. (1996), *Nature* 383: 269–72.
9. A useful review by one of the leading researchers in the field is Kouzarides, T. (2007) *Cell* 128: 693–705.
10. Jenuwein and Allis (2001), *Science* 293: 1074–80.
11. Ng et al. (2010), *Nat Genet.* 42: 790–3.
12. Laumonnier et al. (2005), *J Med Genet.* 42: 780–6.

Chapter 5

1. Fraga et al. (2005), *Proc Natl Acad Sci. USA* 102: 10604–9.
2. Ollikainen et al. (2010), *Human Molecular Genetics* 19: 4176–88.
3. http://www.pbs.org/wgbh/evolution/library/04/4/l_044_02.html
4. http://www.evolutionpages.com/Mouse%20genome%20home.htm
5. Gartner, K. (1990), *Lab Animal* 24:71–7.
6. Whitelaw et al. (2010), *Genome Biology*.
7. Tobi et al. (2009), *HMG*.
8. Kaminen-Ahola et al. (2010).

Chapter 6

1. If you want to know more, try Arthur Koestler's highly readable though exceptionally partisan book, *The Case of the Midwife Toad*.

2. Lumey et al. (1995), *Eur J Obstet Reprod Biol.* 61: 23–20.

3. Lumey (1998), *Proceedings of the Nutrition Society* 57: 129–135.

4. Kaati et al. (2002), *EJHG* 10: 682–688.

5. Morgan et al. (1999), *Nature* 23: 314–8.

6. Wolff et al. (1998), *FASEB J* 12: 949–957.

7. Rakyan et al. (2003), *PNAS* 100: 2538–2543.

8. World Cancer Research Fund figures http://tinyurl.com/47uosv4

9. www.nhs.uk

10. Waterland et al. (2007), *FASEB J* 21: 3380–3385.

11. Ng et al. (2010), *Nature* 467: 963–966.

12. Carone et al. (2010) *Cell* 143: 1084–1096.

13. Anway et al. (2005) *Science* 308: 1466–1469.

14. Guerrero-Bosagna et al. (2010), *PLoS One*: 5.

Chapter 7

1. Surani, Barton and Norris (1984), *Nature* 308: 548–550.

2. Barton, Surani and Norris (1984), *Nature* 311: 374–376.

3. Surani, Barton and Norris (1987), *Nature* 326: 395–397.

4. McGrath and Solter (1984), *Cell* 37: 179–183.

5. Cattanach and Kirk (1985), *Nature* 315: 496–498.

6. Hammoud et al. (2009) *Nature* 460: 473–478.

7. Reik et al. (1987), *Nature* 328: 248–251.

8. Sapienza et al. (1987), *Nature* 328: 251–254.

9. Rakyan et al. (2003), *PNAS* 100: 2538–2543.

Chapter 8

1. Surani, Barton and Norris (1984), *Nature* 308: 548–550.

2. Barton, Surani and Norris (1984), *Nature* 311: 374–376.

3. Surani, Barton and Norris (1987), *Nature* 326: 395–397.

4. Cattanach and Kirk (1985), *Nature* 315: 496–8.

5. De Chiara et al. (1991), *Cell* 64: 845–859.

6. Barlow et al. (1991), *Nature* 349: 84–87.

7. Reviewed in Butler (2009), *Journal of Assisted Reproduction and Genetics*: 477–486

8. Prader, A., Labhart, A. and Willi, H. (1956), *Schweiz Med Wschr*. 86: 1260–1261.

9. http://www.ncbi.nlm.nih.gov/omim/176270

10. Angelman, H. (1965), 'Puppet children': a report of three cases. *Dev Med Child Neurol*. 7: 681–688.

11. http://www.ncbi.nlm.nih.gov/omim/105830

12. Knoll et al.(1989), *American Journal of Medical Genetics* 32: 285–290.

13. Nicholls et al. (1989), *Nature* 342: 281–185.

14. Malcolm et al. (1991), *The Lancet* 337: 694–697.

15. Wiedemann (1964), *J Genet Hum*. 13: 223.

16. Beckwith (1969), *Birth Defects* 5: 188.

17. http://www.ncbi.nlm.nih.gov/omim/130650

18. Silver et al. (1953), *Pediatrics* 12: 368–376.

19. Russell (1954), *Proc Royal Soc Medicine*. 47: 1040–1044.

20. http://www.ncbi.nlm.nih.gov/omim/180860

21. For a useful review, see Gabory et al. (2010), *BioEssays* 32: 473–480.

22. Frost & Moore (2010), *PLoS Genetics* 6 e1001015.

23. Ohinata et al. (2005), *Nature* 436: 207–213.

24. Buiting et al. (2003), *American Journal of Medical Genetics* 72: 571–577.

25. Hammoud et al. (2009), *Nature* 460: 473–478.

26. Ooi et al. (2007), *Nature* 448: 714–717.

27. Stadtfeld et al. (2010), *Nature* 465: 175–81.

28. See Butler (2009), *J Assist Reprod Genet*. 26: 477–486 for a useful review.

29. Kono et al. (2004), *Nature* 428: 860–864.

30. Blewitt et al. (2006), *PLoS Genetics* 2: 399–405.

31. See, for example, http://www.guardian.co.uk/uk/2010/aug/04/cloned-meat-british-bulls-fsa?INTCMP=SRCH

32. For a recent review, see Bukulmez, O. (2009) *Curr Opin Obstet Gynecol*. 21: 260–4.

Chapter 9

1. Jäger et al. (1990), *Nature* 348: 452–4.
2. Margarit et al. (2000), *American Journal of Medical Genetics* 90: 25–8.
3. For a good review of this, see Graves (2010), *Placenta, Supplement A Trophoblast Research* 24: S27–S32.
4. Lyon, M. F. (1961), *Nature* 190: 372–373.
5. Lyon, M. F. (1962), *American Journal of Human Genetics* 14: 135–148.
6. For a useful review, see Okamoto and Heard (2009), *Chromosome Res.* 17: 659–69.
7. McGrath and Solter (1984), *Cell* 37: 179–83.
8. Cattanach and Isaacson (1967), *Genetics* 57: 231–246.
9. Rastan and Robertson (1985), *J Embryol Exp Morphol.* 90: 379–88.
10. Brown et al. (1991), *Nature* 349: 38–44.
11. Borsani et al. (1991), *Nature* 351: 325–329.
12. Brown et al. (1992), *Cell* 71: 527–542.
13. Brockdorff et al. (1992), *Cell* 71: 515–526.
14. Borsani et al. (1991), *Nature* 351: 325–329.
15. For a good review, see Lee, J. T. (2010) *Cold Spring Harbor Perspectives in Biology* 2 a003749.
16. Lee et al. (1996), *Cell* 86: 83–84.
17. Xu et al. (2006), *Science* 311: 1149–52.
18. Lee et al. (1999), *Nature Genetics* 21: 400–404.
19. Navarro et al. (2008), *Science* 321: 1693–1695.
20. Maherali et al. (2007), *Cell Stem Cell* 1: 55–70.
21. Zonana et al. (1993), *Amer J Human Genetics* 52: 78–84.
22. http://www.ncbi.nlm.nih.gov/omim/305100
23. Reviewed in Pinto et al. (2010), *Orphanet Journal of Rare Diseases* 5: 14–23.
24. http://www.ncbi.nlm.nih.gov/omim/310200
25. Pena et al. (1987), *J Neurol Sci.* 79: 337–344.
26. Gordon (2004), *Science* 306: 496–499.
27. For a good review of this, see Graves (2010), *Placenta Supplement A Trophoblast Research* 24: S27–S32.
28. Rao et al. (1997), *Nature Genetics* 16: 54–63.

Chapter 10

1. From *Scientific Autobiography and Other Papers* (1950).
2. Mulder et al. (1975), *Cold Spring Harb Symp Quant Biol.* 39: 397–400.
3. Ohno (1972), *Brookhaven Symposia in Biology* 23: 366–370.
4. See Orgel and Crick (1980), *Nature* 284: 604–607.
5. See Doolittle and Sapienza (1980), *Nature* 284: 601–603.
6. Mattick (2009), *Annals N Y Acad Sci.* 1178: 29–46.
7. http://genome.wellcome.ac.uk/node30006.html
8. http://wiki.wormbase.org/index.php/WS205
9. For a useful review see Qureshi et al. (2010), *Brain Research* 1338: 20–35.
10. Clark and Mattick (2011), *Seminars in Cell and Developmental Biology*, in press at time of publication.
11. Carninci et al. (2005), *Science* 309: 1559–1563.
12. Nagano et al. (2008), *Science* 322: 1717–1720.
13. Zhao et al. (2010), *Molecular Cell* 40: 939–953.
14. Garber et al. (1983), *EMBO J.* 2: 2027–36.
15. Rinn et al. (2007), *Cell* 129: 1311–1323.
16. Ørom et al. (2010), *Cell* 143: 46–58.
17. Lee et al. (1993), *Cell* 75: 843–854.
18. Wightman et al. (1993), *Cell* 75: 858–62.
19. For a good review, see Bartel (2009), *Cell* 136: 215–233.
20. Mattick, J. S. (2010), *BioEssays* 32: 548–552.
21. *Chimpanzee Sequencing and Analysis Consortium* (2005), *Nature* 437: 69–87.
22. Athanasiasdis et al. (2004), *PLoS Biol.* 2: e391.
23. Paz-Yaacov et al. (2010), *Proc Natl Acad Sci. USA* 107: 12174–9.
24. Melton et al. (2010), *Nature* 463: 621–628.
25. Yu et al. (2007), *Science* 318: 1917–20.
26. Marson et al. (2008), *Cell* 134: 521–33.
27. Judson et al. (2009), *Nature Biotechnology* 27: 459–461.
28. Reviewed in Pauli et al. (2011), *Nature Reviews Genetics* 12: 136–149.
29. Giraldez et al. (2006), *Science* 312: 75–79.
30. West et al. (2009), *Nature* 460: 909–913.
31. Vagin et al. (2009), *Genes Dev.* 23: 1749–62.

32. Deng and Lin (2002), *Developmental Cell* 2: 819–830.

33. Aravin et al. (2008), *Molecular Cell* 31: 785–799.

34. Kuramochi-Miyagawa et al. (2008), *Genes and Development* 22: 908–917.

35. Reviewed in Mattick et al. (2009), *BioEssays* 31: 51–59.

36. Wagner et al. (2008), *Dev Cell.* 14: 962–9.

37. Lewejohann et al. (2004), *Behav Brain Res.* 154: 273–89.

38. Clop et al. (2006), *Nature Genetics* 38: 813–818.

39. Abelson et al. (2005), *Science* 310: 317–320.

40. http://www.ncbi.nlm.nih.gov/omim/188400

41. Strak et al. (2008), *Nature Genetics* 40: 751–760.

42. Calin et al. (2004), *Proc Nat Acad Sci. USA* 101: 2999–3004.

43. Volinia et al. (2006), *Proc Natl Acad Sci. USA* 103: 2257–2261.

44. For a useful review see Garzon et al. (2010), *Nature Reviews Drug Discovery* 9: 775–789.

45. Melo et al. (2009), *Nature Genetics* 41: 365–370.

Chapter 11

1. Karon et al. (1973), *Blood* 42: 359–65.

2. Constantinides et al. (1977), *Nature* 267: 364–366.

3. Taylor and Jones (1979), *Cell* 17: 771–779.

4. Jones (2011), *Nature Cell Biology* 13: 2.

5. Jones and Taylor (1980), *Cell* 20: 85–93.

6. Santi et al. (1983), *Cell* 33: 9–10.

7. Ghoshal et al. (2005), *Molecular and Cellular Biology* 25: 4727–4741.

8. Kuo et al. (2007), *Cancer Research* 67: 8248–8254.

9. For an excellent history of the development of SAHA, see Marks and Breslow (2007), *Nature Biotechnology* 25: 84–90.

10. Friend et al. (1971), *Proc Natl Acad Sci. USA* 68: 378–382.

11. Richon et al. (1996), *Proc Natl Acad Sci. USA* 93: 5705–5708.

12. Yoshida et al. (1990), *Journal of Biological Chemistry* 265: 17174–17179.

13. Richon et al. (1998), *Proc Natl Acad Sci. USA* 95: 3003–3007.

14. Herman et al. (1994), *Proc Natl Acad Sci. USA* 91: 9700–9704.

15. Esteller et al. (2000), *Journal of the National Cancer Institute* 92: 564–569.

16. Toyota et al. (1999), *Proc Natl Acad Sci. USA* 96: 8681–8686.

17. Lu et al. (2006), *Oncogene* 25: 230–9.

18. Gery et al. (2007), *Clin Cancer Res.* 13: 1399–404.

19. For a recent review of a disorder where gene therapy is proving broadly effective see Ferrua et al. (2010), *Curr Opin Allergy Clin Immunol.* 10: 551–6.

20. Kantarjian et al. (2006), *Cancer* 106: 1794–1803.

21. Silverman et al. (2002), *J Clin Oncol.* 20: 2429–2440.

22. Duvic et al. (2007), *Blood* 109: 31–39.

23. www.cancer.gov/clinicaltrials/search/results?protocolsearchid=8828355

24. www.lifesciencesworld.com/news/view/11080

25. http://www.masshightech.com/stories/2008/04/21/story1-Epigenetics-is-the-word-on-bio-investors-lips.html

26. Viré et al. (2006), *Nature* 439: 871–874.

27. Schlesinger et al. (2007), *Nature Genetics* 39: 232–236.

28. Shi et al. (2004), *Cell* 29: 119; 941–53.

29. Ooi et al. (2007), *Nature* 448: 714–717.

30. Bachmann et al. (2006), *J Clin Oncology* 24: 268–273.

31. Lim et al. (2010), *Carcinogenesis* 31: 512–20.

32. Kondo et al. (2008), *Nature Genetics* 40: 741–750.

33. Widschwendter et al. (2007), *Nature Genetics* 39: 157–158.

34. Taby and Issa (2010), *CA Cancer J Clin.* 60: 376–92.

35. Bernstein et al. (2006), *Cell* 125: 315–326.

36. Ohm et al. (2007), *Nature Genetics* 39: 237–242.

37. Fabbri et al. (2007), *Proc Natl Acad Sci. USA* 104: 15805–10.

Chapter 12

1. For a recent review, see Heim et al. (2010), *Dev Psychobiol.* 52: 671–90.

2. Yehuda et al. (2001), *Dev Psychopathol.* 13: 733–53.

3. Heim et al. (2000), *JAMA* 284: 592–7.

4. Lee et al. (2005), *Am J Psychiatry* 162: 995–997.

5. Carpenter et al. (2004), *Neuropsychopharm.* 29: 777–784.

6. Weaver et al. (2004), *Nature Neuroscience* 7: 847–854.

7. Murgatroyd et al. (2009), *Nature Neuroscience* 12: 1559–1565.

8. Skene et al. (2010), *Mol Cell* 37: 457–68.

9. McGowan et al. (2009), *Nature Neuroscience* 12: 342–248.

10. http://www.who.int/mental_health/management/depression/definition/en/

11. Reviewed in Uchida et al. (2011), *Neuron* 69: 359–372.

12. Uchida et al. (2011), *Neuron* 69: 359–372.

13. Elliott et al. (2010), *Nature Neuroscience* 13: 1351–1353.

14. Uchida et al. (2011), *Neuron* 69: 359–372.

15. For a useful review of animal models of depression, see Nestler and Hyman (2010), *Nature Neuroscience* 13: 1161–1169.

16. Uchida et al. (2011), *Neuron* 69: 359–372.

17. Weaver et al. (2004), *Nature Neuroscience* 7: 847–854.

18. See, for example, interviews in Buchen (2010), *Nature* 467: 146–148.

19. Mayer et al. (2000), *Nature* 403: 501–502.

20. Tahiliani et al. (2009), *Science* 324: 30–5.

21. Globisch et al. (2010), *PLoS One* 5: e15367.

22. For a useful review of DNA methylation and memory formation, see Day and Sweatt (2010), *Nature Neuroscience* 13: 1319–1329.

23. Korzus et al. (2004), *Neuron* 42: 961–972.

24. Alarcón et al. (2004), *Neuron* 42: 947–959.

25. MacDonald and Roskams (2008), *Dev Dyn.* 237: 2256–2267.

26. Guan et al. (2009), *Nature* 459: 55–60.

27. Fischer et al. (2007), *Nature* 447: 178–182.

28. Im et al. (2010), *Nature Neuroscience* 13: 1120–1127.

29. Deng et al. (2010), *Nature Neuroscience* 13: 1128–1136.

30. Garfield et al. (2011), *Nature* 469: 534–538.

Chapter 13

1. http://www.isaps.org/uploads/news_pdf/Raw_data_Survey2009.pdf

2. Aubert and Lansdorp (2008), *Physiological Reviews* 88: 557–579.

3. For a review of this and other surveys on public attitudes to lifespan extension, see Partridge et al. (2010), *EMBO Reports* 11: 735–737.

4. Bjornsson et al. (2008), *Journal of the American Medical Association* 299: 2877–2883.

5. Gaudet et al. (2003), *Science* 300: 488–492.

6. Eden et al. (2003), *Science* 300: 455.

7.   For a useful review of changes in epigenetic modifications during ageing, see Calvanese et al. (2009), *Ageing Research Reviews* 8: 269–276.

8.   Kennedy et al. (1995), *Cell* 80: 485–496.

9.   Kaeberlein et al. (1999), *Genes and Development* 13: 2570–2580.

10. Dang et al. (2009), *Nature* 459: 802–807.

11. Tissenbaum and Guarente (2001), *Nature* 410: 227–230.

12. Rogina and Helfand (2004), *Proceedings of the National Academy of Sciences USA* 101: 15998–16003.

13. Michishita et al. (2008), *Nature* 452: 492–496.

14. Kawahara et al. (2009), *Cell* 136: 62–74.

15. http://www.ncbi.nlm.nih.gov/omim/277700

16. Michishita et al. (2008), *Nature* 452: 492–496.

17. McCay et al. (1935), *Nutrition* 5: 155–71.

18. Reviewed in Kaeberlein and Powers (2007), *Ageing Research Reviews* 6: 128–140.

19. Partridge et al. (2010), *EMBO Reports* 11: 735–737.

20. Howitz et al. (2003), *Nature* 425: 191–196.

21. Wood et al. (2004), *Nature* 430: 686–689.

22. Baur et al. (2006), *Nature* 444: 337–342.

23. Howitz et al. (2003), *Nature* 425: 191–196.

24. Beher et al. (2009), *Chem Biol Drug Des.* 74: 619–24.

25. Pacholec et al. (2010), *J Biol Chem.* 285: 8340–51.

26. For a review, see Chaturvedi et al. (2010), *Chem Soc Rev.* 39: 435–54.

27. http://www.fiercebiotech.com/story/weak-efficacy-renal-risks-force-gsk-dump-resveratrol-program/2010-12-01

Chapter 14

1.   McCay et al. (1935), *Nutrition* 5: 155–71.

2.   For a useful review of the differences, see Chittka and Chittka (2010), *PLoS Biology* 8: e1000532.

3.   For a useful summary of these processes, see Maleszka (2008), *Epigenetics* 3: 188–192.

4.   Honeybee Genome Sequencing Consortium (2006), *Nature* 443: 931–49.

5.   Wang et al. (2006), *Science* 314: 645–647.

6.   Kucharski et al. (2008), *Science* 319: 1827–1830.

7.   Lyko et al. (2010), *PLos Biol.* 8: e1000506.

8.   Spannhoff et al. (2011), *EMBO Reports* 12: 238–243.

9.   Lockett et al. (2010), *NeuroReport* 21: 812–816.

10. Hunt et al. (2010), *Genome Biol Evol* 2: 719–728.

11. Lyko et al. (2010), *PLos Biol.* 8: e1000506.

12. Bonasio et al. (2010), *Science* 329: 1068–1071

13. Izuta et al. (2009), *Evid Based Complement Alternat Med.* 6: 489–94.

Chapter 15

1.   For a useful review, see Dennis and Peacock (2009), *J Biol* 8: article 57.

2.   For a useful summary of the epigenetic control of vernalisation, see Ahmad et al. (2010), *Molecular Plant* 4: 719–728.

3.   Pien et al. (2008), *Plant Cell* 20: 580–588.

4.   Sung and Amasino (2004), *Nature* 427: 159–164.

5.   De Lucia et al. (2008), *Proc Natl Acad Sci. USA* 105: 16831–16836.

6.   Heo and Sung (2011), *Science* 331: 76–79.

7.   Pant et al. (2008), *Plant J* 53: 731–738.

8.   Palauqui et al. (1997), *EMBO J* 16: 4738–4745

9.   See, for example, Schubert et al. (2006), *EMBO J* 25: 4638–4649.

10. Gehring et al. (2009), *Science* 324: 1447–1451.

11. Hsieh et al. (2009), *Science* 324: 1451–1454

12. Mosher et al. (2009), *Nature* 460: 283–286.

13. Slotkin et al. (2009), *Cell* 136: 461–472.

14. Garnier et al. (2008), *Epigenetics* 3: 14–20.

15. Reviewed in Zhang et al. (2010), *J Genet and Genomics* 37: 1–12.

16. Chan et al. (2005), *Nature Reviews Genetics* 6: 351–360.

17. Cokus et al. (2008), *Nature* 452: 215–219.

Chapter 16

1.   For a recent review, see Wapstra and Warner (2010), *Sex Dev.* 4: 110–8.

2.   Miller et al. (2004), *Fertil Steril.* 81: 954–64.

3.   Watson, J. D. and Crick, F. H. C. (1953), *Nature* 171: 737–738.

4.   Cattanach and Isaacson (1967), *Genetics* 57: 231–246.

5.   For further information, see http://www.sienabiotech.com.

6.   Mack, G. S. (2010), *Nat Biotechnol.* 28: 1259–66.

7.   MRC Vitamin Study Research Group (1991), *Lancet* 338: 131–7.

8.   Waterland et al. (2006), *Hum Mol Genet.* 15: 705–16.

9.   Reviewed in Calvanese et al. (2009), *Ageing Research Reviews* 8: 268–276.

10.   Reviewed in Guilloteau et al. (2010), *Res Rev.* 23: 366–84.

11.   Palsdottir et al. (2008), *PLoS Genet.* June 20, 4: e1000099.

12.   Abstract from Palsdottir et al. (2010), *Wellcome Trust Conference on Signalling to Chromatin Hinxton UK*

13.   See for example Okabe et al. (1995), *Biol PharmBull.* 18: 1665–70.

14.   Nakahata et al. (2008), *Cell* 134: 329–40.

15.   Katada et al. (2010), *Nat Struct Mol Biol.* 17: 1414–21.

16.   Gregoretti et al. (2004), *J Mol Biol.* 338: 17–31.

Glossary

Autosomes   The chromosomes which are not sex chromosomes. There are 22 pairs of autosomes in humans.

Blastocyst   Very early mammalian embryo, consisting of about 100 cells. The blastocyst comprises a hollow ball of cells that will give rise to the placenta, surrounding a smaller, denser ball of cells that will give rise to the body of the embryo .

Chromatin   DNA in combination with its associated proteins, especially histone proteins.

Concordance   The degree to which two genetically identical individuals are identical phenotypically.

CpG   A cytosine nucleotide followed by a guanine nucleotide in DNA. CpG motifs can undergo methylation on the C.

Discordance   The degree to which two genetically identical individuals are non-identical phenotypically.

DNA replication   Copying DNA to create new DNA molecules which are identical to the original.

DNMT   DNA methyltransferase. An enzyme that can add methyl groups to cytosine bases in DNA.

Epigenome   All the epigenetic modifications on the DNA genome and its associated histone proteins.

ES Cells   Embryonic stem cells. Pluripotent cells experimentally derived from the Inner Cell Mass.

Exon   Region of a gene that codes for a section that is present in the final version of the mRNA copied from that gene. Most, but not all, exons encode amino acids in the final protein produced from a gene.

Gamete   An egg or a sperm.

Genome   All the DNA in the nucleus of a cell.

Germline   The cells that pass on genetic information from parent to child. These are the eggs and the sperm (and their precursors).

HDAC   Histone deacetylase. An enzyme that can remove acetyl groups from histone proteins.

Histones   Globular proteins that are closely associated with DNA, and which can be epigenetically modified.

Imprinting   Phenomenon in which expression of certain genes depends on whether they were inherited from the mother or the father.

Inner Cell Mass (ICM)   The pluripotent cells in the inside of the early blastocyst that will give rise to all the cells of the body.

Intron   Region of a gene that codes for a section that is removed from the final version of the mRNA copied from that gene.

iPS Cells   Induced pluripotent stem cells. Produced by reprogramming mature cells with specific genes that cause terminally differentiated cells to revert into pluripotent ones.

kb   Kilobase. 1,000 base pairs.

miRNA   Micro RNA. Small RNA molecules that are copied from DNA but that don't code for proteins. miRNAs are a subset of ncRNAs

mRNA   Messenger RNA. Copied from DNA and codes for proteins.

ncRNA   Non-coding RNA. Copied from DNA and doesn't code for proteins.

MZ Twins   Monozygotic/ identical twins, formed when an early embryo splits in two.

Neurotransmitter   A chemical produced by one brain cell that acts on another brain cell to alter its behaviour.

Nucleosome   Combination of eight specific histone molecules with DNA wrapped around them.

Phenotype   The observable characteristics or traits of an organism.

Pluripotency The ability of a cell to give rise to most other cell types. Typically, pluripotent mammalian cells give rise to all cells of the body, but not the cells of the placenta.

Priomordial germ cells Very specialised cells created in early development, which give rise ultimately to the gametes.

Promoter Region in front of a gene that controls how a gene is switched on.

Pronucleus The nucleus of a sperm or egg, following entry of a sperm into an egg, but before the two nuclei fuse.

Retrotransposons Unusual segments of DNA that don't code for protein and can move between different locations in the genome. Believed to have originated from viruses.

RNA Single stranded copy of a specific region of DNA. The term RNA stands for ribonucleic acid. It includes various different classes of RNA molecules including miRNA, mRNA and ncRNA.

Sex chromosomes The X and Y chromosomes that govern sex determination in mammals. Normally, females have two X chromosomes and males have one X and one Y chromosome.

Somatic cells Cells of the body.

Somatic Cell Nuclear Transfer (SCNT) The transfer of the nucleus from a mature cell into another cell, usually an egg.

Somatic mutations Mutations that happen in the cells of the body, rather than ones that have been inherited via sperm or eggs.

Stochastic variation A random change or fluctuation.

Totipotency The ability of a cell to give rise to all cells of the body and the placenta.

Transcription Copying DNA to create RNA molecules.

Transgenerational inheritance The phenomenon in which phenotypic changes in one generation are passed on to the next, without any alteration in the genetic code.

Uniparental disomy A situation where both members of a pair of chromosomes have been inherited from one parent, rather than one from each parent. For example, maternal uniparental disomy of chromosome 11 would mean both copies of chromosome 11 had come from the mother.

Vernalisation   The process where plants need a period of cold before they will flower.

Zygote   The totipotent cell formed when an egg and a sperm fuse.

Index

Junk DNA

A Journey Through the Dark Matter of the Genome

Nessa Carey's new book *Junk DNA* is a thrilling exploration of the cutting edge of human science.

For decades after the structure of DNA was identified, scientists focused purely on genes, the regions of the genome that contain codes for the production of proteins. Other regions – 98% of the human genome – were dismissed as 'junk'. But in recent years researchers have discovered that variations in this 'junk' DNA underlie many previously intractable diseases, and they can now generate new approaches to tackling them.

Nessa Carey explores, for the first time for a general audience, how junk DNA plays an important role in areas as diverse as genetic diseases, viral infections, sex determination, human biological complexity, disease treatments, even evolution itself.

ISBN: 9781848318267 (hardback)
9781848319158 (paperback)
9781848318274 (ebook)